Bergli Books

MONEY LOGGING

ON THE TRAIL OF THE ASIAN TIMBER MAFIA

BY LUKAS STRAUMANN

First published in German as *Raubzug auf den Regenwald. Auf den Spuren der malaysischen Holzmafia* Copyright © 2014 Salis Verlag AG, Zurich
This book was made possible through the generous support of the Bruno Manser Fund.
Text Copyright © 2014 Lukas Straumann, all rights reserved
Book Copyright © 2014 Bergli Books, an imprint of Schwabe AG, Basel, Switzerland
Maps and Tables Copyright © 2014 Salis Verlag AG, Zurich
Cover design: Christoph Lanz, moxi ltd, Biel, Switzerland
Illustrations: Daniela Trunk, Zug, Switzerland
Tables: Johanna Michel, Bern, Switzerland
Printed by: Schwabe AG, Muttenz/Basel, Switzerland
Printed in Switzerland
ISBN 978-3-905252-68-2
ISBN E-Book (EPUB) 978-3-905252-69-9
ISBN E-Book (PDF) 978-3-905252-70-5
ISBN E-Book (Mobipocket) 978-3-905252-71-2

www.bergli.ch

TABLE OF CONTENTS

IN MEMORY OF BRUNO MANSER
(1954–2000)

Hedda Morrison

Penan man encountered in

the deep jungle c.1950

Gelatin silver photograph

National Gallery of Australia,

Canberra

Bequest of Hedda Morrison

1992

FOREWORD

By Mutang Urud

I was born in a village in the "Heart of Borneo" as Tom Harrison described it, near the remote headwaters of the Limbang River, in the Malaysian state of Sarawak. There is nothing more beautiful than the rainforests of Borneo where I spent my childhood. It was both our playground and our sweets shop. We foraged for *rinuan* honey and ground fruits on the forest floor, and climbed up vines and fruit trees to feed our sugar-starved young souls. Growing up surrounded by mountains, the forest was our only world, and under the dark canopy where the noonday seems like dusk, only the calls of birds and cicadas told us the time of day. Borneo's virgin forest is also home to tens of thousands of insects, hundreds of bird species, and many mammals that are found nowhere else. A single hectare of our forest supports more tree species than all of Europe.

As a young adult in the 1970s, I watched the loggers not only destroy the forest, but divide communities with corrupting bribes and pay-offs. They were like thieves in the night; indeed, they were working in such haste that their machinery could be heard at midnight, even on Sunday. Our ancestral land has been desecrated, our history erased, the very memory of our origins lost. As a young idealist, I could not stand by while this crime was occurring. In the late 1980s, I helped organise blockades to stop the bulldozers and chainsaws. I founded the Sarawak Indigenous Peoples' Alliance as a rallying point for our peoples' resistance. Only reluctantly did I travel to twenty-five cities in thirteen countries to tell the world what was happening to our homeland. Back in Sarawak, police attacked our blockades and sent many people to jail. I was arrested, interrogated, and held in solitary confinement. Upon my release, I left Malaysia to speak about these environmental crimes at the Earth Summit in Rio de Janeiro. In 1992, I addressed the United Nations General Assembly in New York in support of land rights for indigenous peoples. Unable to return home, I studied anthropology in Canada in order to acquire new skills that would help me save some of what was being lost.

Fearing arrest, for twenty years I dared not return to my homeland. When I finally did, I found that the ecological crimes had only increased. The forest I had loved was almost gone. Rainforests that had been the home of human beings for at least 40,000 years had been destroyed in little more than thirty. Close to 90% of Sarawak's ancient forest is now gone. Only 11% of the primary growth remains. How did it disappear?

I applaud my dear colleague Lukas Straumann for his diligence and investigative skill in writing the book that follows. His research exposes the wanton greed that has fuelled the destruction of the place I call home.

This book investigates two crimes. The first is how a single man, Abdul Taib Mahmud, along with a small group of very rich politicians and businessmen could destroy the richest ecosystem on earth despite not owning it, despite local and global outcry, despite international laws and regulations. Simply put: Who has stolen our trees?

The second crime is more subtle. Surely, if my people have lost their ecosystem, their traditional way of life, their clean drinking water, and their freedom to roam the forests, they must have gained something. Yet they haven't. Many of the people of Sarawak are as poor as they were when I was born. And yet, the value of the trees that have been felled is estimated to exceed US$50 billion. This profit has fed corruption, kept oligarchs in power, been used to commit further crimes. Fortunes have moved through the world's financial system, mostly secretly, to places as distant as Zurich, London, Sydney, San Francisco, and Ottawa.

Lukas Straumann shows how this, one of the greatest environmental crimes in history, is much bigger than just the theft of trees. It is also about power, more precisely, how a corrupt autocrat has liquidated a forest in order to keep himself at the helm of a state. For my people it is also more than a question of trees. It is about our culture they have stolen.

This book should be essential reading for anyone who uses a bank, buys property, or invests in the stock market. Only by understanding how a rainforest can be converted into a building as far away as the FBI headquarters in Seattle can we hope to stop the kind of corruption that threatens the world's natural places, and the people for whom these are home.

Mutang Urud
Montreal, Canada
July 2014

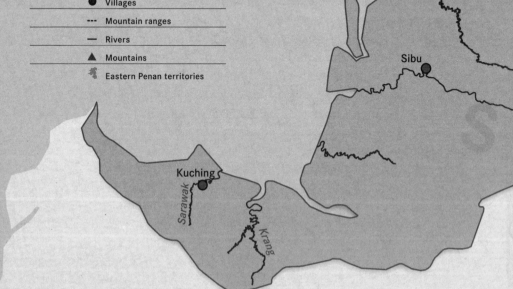

MALAYSIA

SARAWAK

THAILAND

Kuala Lumpur

SINGAPORE

THE
PHILIPPINES

BORNEO

INDONESIA

INDIAN OCEAN

SOUTH

● Cities

● Villages

--- Mountain ranges

— Rivers

▲ Mountains

Eastern Penan territories

Mukah

Sibu

Kuching

Sarawak

Krang

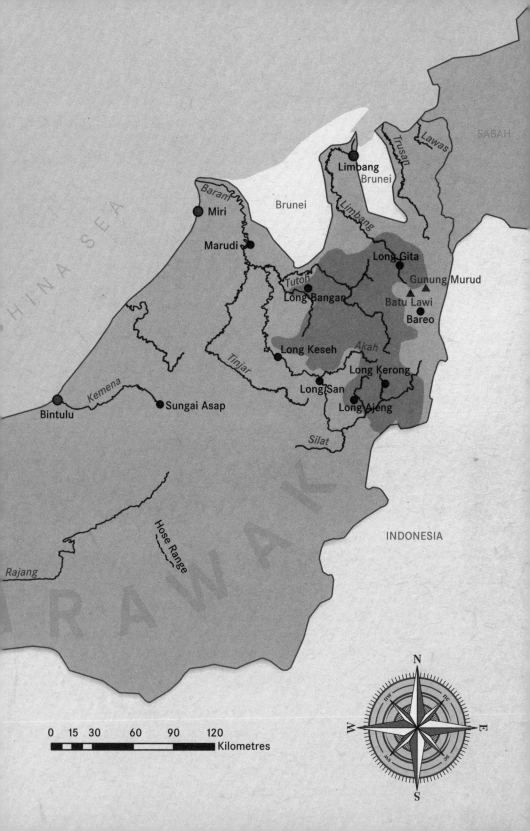

SABAH

CHINA SEA

Baram

Miri

Marudi

Brunei

Limbang
Brunei

Trusan

Lawas

Limbang

Long Gita

Gunung Murud

Tutoh

Long Bangan

Batu Lawi

Bareo

Long Keseh

Akah

Long Kerong

Tinjar

Long San

Long Ajeng

Kemena

Bintulu

Sungai Asap

Silat

Hose Range

INDONESIA

Rajang

SARAWAK

| 0 | 15 | 30 | 60 | 90 | 120 |
Kilometres

N

NW NE

W   E

SW SE

S

FOLLOW THE MONEY

An insider tells all: Rainforest despot Taib has amassed a worldwide real estate empire worth hundreds of millions of dollars. Even the FBI is one of Taib's tenants. The nerve centre of the property empire is in an upmarket suburb of the Canadian capital, Ottawa. A secret rendezvous with the whistle-blower ends in a nightmare.

TAIB'S SECRET REAL ESTATE EMPIRE

On 20 June 2010, Clare Rewcastle's Blackberry flashed. A curious message had landed in her inbox: "I was Sulaiman Taib's Chief Operating Officer in the US for twelve years. I have sensitive information and am ready to share it. But are you ready to fight with Taib? Careful, my phones are tapped and my computer is compromised. Ross Boyert."

Four months later, Ross Boyert was dead.

Clare Rewcastle, a former BBC journalist, did not hesitate for long before contacting the Bruno Manser Fund. "We've got to meet Boyert at once," she said to me over the telephone. "This man holds the key to Taib's secret real estate empire. We've got to go over to the US as soon as possible. I never thought we'd find him." Two days later, I was sitting in an aircraft bound for Los Angeles.

Clare Rewcastle lives in London now and is married to a brother of the former British Prime Minister, Gordon Brown, but she spent her childhood in Sarawak, Malaysia, as the daughter of British colonial servants. She left at the age of eight, returning to the United Kingdom with her family. At the end of 2005, she travelled to Sarawak to attend an environmental conference and was shocked to find the country of her childhood unrecognizable. 90% of Sarawak's exploitable timber had been felled. Land that had once been covered in dense rainforests had been replaced by palm oil plantations. The indigenous inhabitants' longhouses were gone, and in their place were the logging companies' camps. The people in the countryside were poorer and worse off than they had been when Clare was a child, but, in stark contrast, the mansions of the leading politicians and timber barons glistened in the towns and cities.

One man had ruled Sarawak for over thirty years: Abdul Taib bin Mahmud, known in Malaysia as "Taib Mahmud" or simply "Taib". With holdings in more than 400 businesses in twenty-five countries and offshore financial centres, Taib's family is a global player. It is estimated that Taib's wealth is worth a total of 15 billion US dollars, making him one of the richest and most powerful men in Southeast Asia.[1] Under Taib's rule,

Sarawak had become a "hotspot" in the global crisis afflicting tropical rainforests.[2]

Clare Rewcastle first visited us at the Bruno Manser Fund in Basel, Switzerland, in 2009, and we agreed to work together to expose the crimes of Taib and his entourage. Early in 2010, the energetic journalist launched her blog *Sarawak Report*, which soon became one of Malaysia's best-read news pages. Together, we scoured the Internet—Clare, from her base in London, and myself, in my office in Basel—searching for information about Taib's global businesses. Very quickly it became clear to us that Taib must have earned billions illegally from the timber trade, and he must have parked that fortune somewhere abroad. But where? If we could find it, we would be one step closer to the smoking gun we needed in our fight for the rainforests of Sarawak. "Follow the money" had become our motto, and now, out of the blue, we were suddenly hot on the trail of Taib's investments abroad.

Ross Boyert's existence was not news to us. We'd heard about him through the Californian NGO The Borneo Project, but all our attempts to track down the whistle-blower had ended in failure. We had not even known whether he was still alive. Until now.

We met Ross Boyert and his wife Rita (name changed) on Wednesday, 23 June 2010, at eight o'clock in the morning in the bar of the Marriott at Los Angeles airport, a high-rise hotel built in the 1970s that was beginning to show signs of age. Clare and I had flown in from Europe the evening before. The Boyerts turned out to be a fashionable pair, both around sixty and both dressed in designer clothes. The strong, dark-haired Ross with his bushy eyebrows greeted us jovially. Rita, too, a graceful blonde woman in a dark dress with a pearl necklace, was visibly pleased to see us. "Don't give us any advance notice of when you're coming and don't call until you're here," Ross had warned us on the telephone. "We'll come to the airport immediately. That's the only way we can meet without being shadowed. Since I initiated proceedings against the Taib family, our life has become hell."

With the introductions completed, we hurriedly withdrew to a meeting room in the Marriott basement, where we would be able to talk without being interrupted. As a final gesture, Ross turned to look anxiously at the hotel entrance, but there was no one there to be seen.

"It's terrible. We're being followed day and night," Rita Boyert burst out the instant the door to the meeting room was closed.

Ross added: "Taib and his people have inflicted the same on us as on the Borneo rainforest: destruction, annihilation, theft, and betrayal. Ruination for the sake of ruination. I see no future any more, and that's precisely what they want."

Always a shrewd journalist, Clare had started recording the conversation. She began asking precise questions. I merely watched and listened.

"Taib owns properties worth 80 million US dollars in San Francisco and Seattle," Ross explained, "and I administered them for twelve years on behalf of his son, Sulaiman. Sakti International Corporation, Wallysons Inc., and W.A. Boylston are companies owned by the Taib family, with properties on the west coast of the USA. The companies are registered in the names of Taib's children and his brothers and sisters, but in reality they belong to him in person. Here's proof."[3]

Ross Boyert put a hand into his leather case and pulled out a sheaf of photocopies. He placed one document in the middle of the wide conference table. "Articles of Incorporation of Sakti Corporation" read the title of the deed creating Taib's Sakti real estate business on 5 March 1987.

Ross flipped through the documents and then snatched a second paper. Its title was "Certificate of Amendment of Articles of Incorporation", and at the bottom was the official seal of the State of California. The document proved that the Sakti Corporation had changed its name to the Sakti International Corporation on 10 September 1987, and that act was witnessed with the neat signatures of the company's directors at the time, Taib's two brothers, Onn and Arip, and the elder of Taib's two sons, Mahmud Abu Bekir, known as Abu Bekir.

"But here's the real proof," said Ross. He stood and pointed triumphantly at a two-page document dated 8 April 1988 with the cumbersome title of "Action by Unanimous Written Consent of the Board of Directors of Sakti International Corporation". The document reported the issuing of one thousand Sakti shares at one dollar per share, split unequally between five people: Taib's two brothers: Onn and Arip; and three of Taib's children: Abu Bekir, Jamilah, and Sulaiman Abdul Rahman.

"All the shares are formally held by Taib's brothers and children," Ross Boyert explained, "but the trick is that half the shares are held in trust for Taib personally. His name does not appear in the share register, although he is the biggest Sakti shareholder." And, in point of fact, in the column with the heading "Number of Shares", it became clear for whom it was that Taib's brothers and children held the shares: "200 of which to be held in trust for Abdul Taib Mahmud" was the endorsement next to the 400 shares of his brother Onn. In the case of his brother Arip and his two sons, it was 100 shares each, giving Taib a total holding of 500 out of the 1,000 shares being held in trust for him. With the secret 50% shareholding, it is also clear who had control over the company: the chief minister in person and he alone. Here, for the first time, we had proof of the chief minister's secret wealth.

Ross Boyert handed the documents over to Clare and me, and then he sat down again. Suddenly it seemed as if that blazing fire inside him had been snuffed out. He was once again very apprehensive. Slowly, quietly, and haltingly in that windowless cellar meeting room, Ross and Rita Boyert began to relate the story of their life as Taib's confidential agents in the USA.

AMERICAN DREAM

Ross Boyert was born in 1950 and grew up in California in a family with a Polish background. Despite having a tough time in his younger years, Ross completed his studies at the University of Southern California in Los Angeles and wanted to give himself a better life than his parents had known. He chose a safe but potentially lucrative career—accountancy—and went on to specialise in real estate management.

While studying, Ross Boyert shared an apartment with the future film star Kurt Russell, and was at home in a circle of upwardly mobile young people. Hollywood was nearby, with its prosperity, glamour, and a glitzy life full of fun and enjoyment. The American dream seemed within his grasp. He was offered a good job when he turned thirty, and it took him to the oil metropolis of Houston, Texas. It was there, in 1984, that he married Rita Nowak (name changed), who had Polish roots as he did. The couple had one daughter, who was born the following year.

Ross went on to various important positions in real estate in Texas and California. Then, at the end of 1994, when he was in his mid-forties and well-experienced, he joined Taib's Sakti International Corporation at its headquarters in San Francisco. At that time, the company was in serious financial difficulties.

"Taib's son had squandered a huge amount of money in a very short period of time, and the company was on the verge of bankruptcy," Ross told us. "He had had absolutely no experience of business when Taib entrusted him with executing projects of which he hadn't the faintest idea. He was in urgent need of an experienced real estate manager. It was a tailor-made job for me."[4]

Ross was hired by Taib's younger son, Sulaiman Abdul Rahman, who was called Rahman or "Ray" in the USA, whereas he was known as "Sulaiman" at home in Malaysia (in order to avoid any confusion with Taib's uncle Rahman, he is referred to as "Sulaiman" throughout this book). Sulaiman, the Taibs' third child, was born in 1968. At the end of the 1980s, he went to California to study. As a son of the chief minister of Sarawak, he had been born with all the wealth he could ever need, and the ardent automobile enthusiast was determined to enjoy the American way of life to the full.

One of Sulaiman's student friends from the Philippines, looking back on those days, wrote that Sulaiman "was the first person I knew that had personally owned a vast number of ultra exotic automobiles. He seemingly had a new car every couple of weeks, and had a dedicated car shop and storage facility that catered to his every whim. He had everything from an old K-type Mercedes Benz worth over a million dollars, an SL Gullwing, Ferrari 355 Spider, Rolls Royce Corniche, Maserati Kamsin, to a 'regular' S500 Mercedes. He had so many cars that he would routinely send them off to his home in Sarawak to clear his garage of the clutter."[5] The student from Sarawak was most definitely not short of money. An acquaintance, who stole a secret glance inside Sulaiman's bank book one day, reports seeing the sum of four million dollars—presumably pocket money from his father, Taib.

In 1991, the 23-year-old Sulaiman married the 20-year-old Elisa (later Anisa) Chan, daughter of George Chan, a Sarawak politician of Chinese origin, who was soon to become one of Taib's most important political allies and would even ascend to the rank of Taib's deputy.[6] The matrimonial bond within Sarawak's political establishment was celebrated as the marriage of the year, with 7,000 roses and 20,000 guests. The couple later went on to have four children. Sarawak's press, with its loyal, pro-government line, reported on the wedding at epic length and published photographs showing the newly-wed couple beaming in front of a wedding cake several metres high.

Playboy Sulaiman, however, had a dark side to him too. "Once, in a fit of rage, he wrecked one of his Bugattis, one of the most expensive cars in the world, with a fire extinguisher," Ross Boyert remembers. "I saw the battered sports car with my own eyes. The windscreen and the hood had been smashed up. It was a shocking sight." Later, Sulaiman would beat his wife as well (after a few years she filed for divorce). In 2003, Sulaiman was in the headlines again, this time for beating up his girlfriend—a well-known Malaysian television presenter—so brutally in a bar in Kuala Lumpur that she needed hospital treatment.[7]

In Ross Boyert, who was more than twenty years his senior, the 26-year-old Sulaiman had found a capable and discreet manager for the properties owned by the family on the west coast of the USA. Ross set to work without delay. He began by working from home, but after a few months moved into an office in Sakti's headquarters in the financial district of San Francisco, with the cable cars rattling right past the doorstep. The historic building at 260 California Street had been built not long after the Great Earthquake of 1906, when the whole city still lay in ruins. In 1988, the Taibs had acquired the elegant eleven-storey building for 13 million US dollars.[8]

Shortly after Ross's appointment, Sulaiman left the USA and moved back to Malaysia with his family. He kept in touch with Ross by telephone and fax, and Ross had to report on Sakti's financial results to various cover addresses in Singapore and Kuala Lumpur, always with the strictest confidentiality, so that it would not be possible for anyone to learn any details regarding the ownership of Sakti International.

"Keep up the good work!" Sulaiman wrote cordially to Ross from Malaysia at the end of 1995. He told Ross of the birth of his youngest daughter and sent him photographs from Malaysia, so that he would be able to gain some impression of life in Southeast Asia. At that time, Ross was working incessantly for Sakti, organising mortgages, negotiating with potential tenants, and supervising the renovation work on Taib properties in San Francisco and Seattle that were in need of improvement.

The work for Sakti proved to be profitable for Ross Boyert. In addition to his basic salary of 115,000 US dollars, he received bonuses for the successful negotiation of mortgages and the conclusion of rental contracts for Taib properties.[9] In his most successful business year, 1999, he pocketed "incentive loan fees" worth more than 700,000 US dollars. Everyone was satisfied with his performance.

The Boyert family owed its social ascent to the abundant flow of Taib dollars, and they sought to rub shoulders with America's best and brightest. In the spring of 1999, the Boyerts moved out of San Francisco to the affluent residential district of Atherton, some 50 kilometres to the south. There they purchased a property worth over a million dollars, surrounded by towering trees. Their only daughter was given a horse of her own and sent to an expensive private school. It was not long, however, before the dark clouds began to gather in the Californian sky. The seeds of Ross's demise had already been planted in his greatest successes.

TOP SECRET—THE TAIBS AND THE FBI

Ross Boyert managed his biggest coup at the end of 1998. It so happened that the FBI was urgently in need of new premises for its northwest headquarters. Ross negotiated a long-term lease with the US federal government for the Abraham Lincoln Building, a multi-storey edifice in the centre of Seattle purchased seven years before by Sulaiman Taib, the student, for 17 million dollars[10]—on behalf of his father, the chief minister, according to Ross. More than ten million dollars were needed for refurbishment of the building, and the funds had to be found at once. Ross claimed that at this point, Taib's son promised Ross 50% of the resulting profits if he could complete the renovation work without needing additional capital. It was a casual promise, and Sulaiman never confirmed it in writing, just as he had never given Ross a written employment contract.

The renovation work at 1110 E 3rd Avenue was indeed completed in a year, and the FBI moved in. Ever since then, emblazoned below the man-sized FBI seal, surrounded by thirteen golden stars, stands the motto of America's top crime-fighting organisation: "Fidelity—Bravery—Integrity". Taib could be satisified that Ross had found such respectable tenants for his property.

"As the manager of the FBI building, I needed 'top secret' security clearance before I was allowed into the place," Ross recounted with an evident note of pride in his voice. "After all, Seattle is one of the two FBI locations from which it combats global terrorism." Every time Ross travelled abroad, he had to face questioning by the FBI on his return, and give precise details about the purpose, itinerary, and duration of his stay abroad.

It is unthinkable that the FBI did not know that its new Seattle headquarters belonged to the family of a corrupt Malaysian politician. There is no evidence, however, that this was a source of concern for any of the senior FBI personnel, although the Seattle FBI boasts on its website that one of its priorities is to "combat public corruption at all levels".[11] Taib was probably helped behind the scenes by his ever-improving relationships in the very highest political and business circles.

For Ross, leasing the Abraham Lincoln Building to the FBI was confirmation that Taib's purported corruption and intrigues could not be all that bad after all, and that he had no reason to question his association with the family. And yet, as someone with a keen interest in the world around him, he used the still-young Internet more and more to follow news from Sarawak. He couldn't help but begin to wonder how the Taibs had made so much money, and why everything had to be kept such a closely guarded secret.

"Sulaiman was, of course, a pampered good-for-nothing," Ross said in retrospect. "At heart, however, he was a nice enough young man. And Laila and the rest of the family, who repeatedly came to the USA on vacation, they all seemed to me to be decent people. At that time, I did not yet know what cruelty these people were capable of, and how much abject poverty they had caused. I simply told myself: If the FBI has checked out the Taibs' background and sees no problem in being their tenant, why should I let the matter trouble my conscience?"

The only one of them whom Ross could not stand from the very beginning was Taib's Canadian son-in-law, Sean Murray, the husband of his daughter Jamilah. When the conversation turned to the subject of Sean Murray, Ross's voice took on a bitter tone: "Even the very first time we met, Sean boasted that the Taibs' family fortune was worth over a billion dollars. Later on, however, when I was in urgent need of fresh capital for renovation work, there was no money there to be had, and I had to organise everything myself."

The notes Ross jotted down under the name "Sean Murray" on the occasion of his first interview at Sakti on 8 December 1994 include three telephone numbers: one for the sister company, Sakto Corporation in Ottawa, one for "Residence," and one for "London", where Sean was busily building up a new real estate company.[12] It was three months later, on a mild February Sunday, that Ross first came face-to-face with the sandy-haired 32-year-old Canadian, who had travelled 4,000 kilometres to San Francisco from the icy cold of Ontario.

POLITICS OF MARRIAGE OR "LIFE IS ONLY WHAT YOU MAKE OF IT"

Sean Murray was born in 1963 and grew up in the affluent Ottawa suburb of Rockcliffe Park, as the child of Irish immigrants. Sean's father Tim had immigrated to Canada in 1957 after studying architecture in Dublin and Liverpool. Four years later, he and his brother Pat founded the architectural practice of Murray & Murray in Ottawa.

The two Murray brothers were certainly talented architects and skilful networkers in the Irish-Canadian community, and were soon core members of the Ottawa River establishment. They quickly built up an excellent reputation for themselves and managed to land numerous public contracts: building Ottawa's international airport, renovation work on the headquarters of the time-honoured Royal Canadian Mounted Police, and modernisation of the Canadian Supreme Court, among others. Ottawa's new city hall, the Saudi-Arabian Embassy, and even the Papal altar for John Paul II's visit to Canada in 1984 were designed on the Murrays' drawing boards.[13]

They sent their sons to the smart Ashbury College in Rockcliffe Park. Sean and his brother, Thady, managed to complete their college studies, as did their cousins (Pat's sons): Patrick, Brian, and Christopher. Sean's sister, Sarah, and their cousin, Fiona, meanwhile attended Rockcliffe's equally prestigious girls' school, Elmwood, with its lofty motto "Summa summarum" (the highest of the high). Many of these relatives were later to play an important role in the Taibs' family business.

Rockcliffe Park, perched on wooded hill, is one of the wealthiest residential districts in all of Canada, and, with 2,000 inhabitants, functions like a village. Everyone knows everyone else, and even the mansions of the super-rich boast no sort of protective fence. Embassy residences stand next to the homes of successful business people, such as software baron and aircraft collector Michael Potter, or Michael Cowpland, who made a fortune with the graphics software Corel Draw and built a glass palace in Rockcliffe Park. For 15 years, Sean Murray's uncle Pat was mayor of Rockcliffe Park until the community lost its autonomy in 2001 to become part of Ottawa.

In the early 1980s, a family from Malaysia appeared, seeking good schools for their children—and a safe haven for their flight capital. The head of the family had a long ministerial career in Kuala Lumpur behind him, and had become chief minister of Sarawak. The Taibs and their assets—huge even in those days—were welcomed with open arms in Rockcliffe Park.

FAMILY CIRCLES–TAIB'S FAMILY NETWORK

THE TAIB MAFIA
TIMBER FAMILY *1936
TAIB MAHMUD
NETWORK
Yang Amat Mat Bermohat Pehin Sri Haji Abdul

CHILDREN

MAHMUD ABU BEKIR TAIB *1963 ∞2 Elena Vasilenko

JAMILAH HAMIDAH TAIB MURRAY *1960 ∞ Sean Patrick Murray

HANIFAH HAJAR TAIB-ALSREE *1972 ∞ Syed Ahmad Alsree

SULAIMAN ABDUL RAHMAN TAIB *1968 ∞ Anisa Hamidah Chan

SPOUSES

∞1 LAILA CHALECKI 1941–2009

∞2 RAGAD WALEED ALKURDI *1982

SIBLINGS

ZALEHA MAHMUD *1949 ∞ Salleh Abdullah

AISAH ZAINAB MAHMUD *1947 ∞ Ruhanie Ahmad

MOHAMAD ALI MAHMUD *1950 ∞ Fatimah Mohamed

AMAR HAJAR FREDAHANUM MAHMUD *1953 ∞ Abdul Aziz Haji Husain

RAZIAH *1955 MAHMUD ∞3 Robert Geneid

MOHD TUFAIL MAHMUD *1952 ∞ Suhing Bnjang

ARIP MAHMUD 1957–2005 ∞ 3 Lily Teo

IBRAHIM 1943–2012 MAHMUD ∞ Hayjah Abdullah

MENTOR

ONN MAHMUD *1949 ∞ Halimatun binti Abdul Ghani

Taib's personal

UNCLE

ABDUL RAHMAN YA'KUB *1928

COUSIN COUSINS

FATIMAH ABDUL RAHMAN
NORLIA ABDUL RAHMAN
NORAH ABDUL RAHMAN
ABDUL HAMED SEPAWI *1949

★ Born

∞ Taib's 1st and 2nd wives

✦ Sister / ■ Brother

◣ Female cousins / ◣ male cousins

∞ Married to

Source: BMF 2012; The Taib Timber Mafia

Back in summer 1981, Sean Murray still knew nothing about Malaysia. Could the eighteen-year-old even have found the island of Borneo on a map? What was certain was that after nine years of Ashbury College, the high-school boy had had enough of anything that smacked of school. He much preferred to sit at home and to play with his synthesiser, or jam with friends in his rock band. His only remaining enthusiasm was for the school rugby and ice-hockey teams. His schoolmates called him Clam—perhaps because he was known as uncommunicative.[14] In any case, he wanted nothing more than for school to come to an end. He dreamt of studying to become an engineer—and of finding a rich and beautiful woman.

The woman of his dreams would indeed enter his life very soon. In his last year at school, a new schoolmate joined him there, the 18-year-old Abu Bekir from Sarawak, Malaysia, who attended class 12 A.[15] When he made the move to Ottawa, Abu Bekir was accompanied by his sister, Jamilah, who was three years older than he. She was a self-confident young lady with flowing hair and sparkling, dark eyes. Jamilah, whose beauty no one could deny, had been registered for the final year at Elmwood School, which was only a few hundred metres away from Ashbury College.[16] They told their new classmates that they were descendants of the Malaysian royal family. It did not take long for the extroverted Jamilah to be given the nickname of "the Princess". Many of the boys in Rockcliffe Park soon set their sights on her. In summer 1982, Sean, Jamilah, and Abu Bekir all graduated.[17]

How Sean Murray won over his Malaysian princess is not relevant to this story—the important fact is that Sean Murray gave up his dream of studying to become an engineer, and, instead, applied to study business management at Ottawa's Carleton University, the same subject Jamilah had decided to pursue. To do business together seemed to be the passion of the two young lovers. "Life is only what you make of it" was the motto espoused by Sean. And clearly, he and Jamilah planned to make much of it.

The liaison was extremely propitious for both the Taibs and the Murrays. A potentate's daughter from Borneo joined with the son of an Irish-Canadian property tycoon; fresh capital from the Far East joined with political connections in Ontario. In Rockcliffe Park, this is what dreams are made of.

The lovers' fellow students remember Jamilah giving her darling Sean a red Mercedes convertible so the pair could cruise the streets of Ottawa. But for both it was a serious relationship. Five years after finishing school they married. For Sean, love was a strong enough motive for him to forsake the Irish-Catholic legacy of his forebears. He converted to Islam and it was under the name of Mohammad Nor Hisham Murray that he married the daughter of the chief minister of Sarawak in 1987. From then onwards, his Malaysian in-laws called him Hisham, but Sean Murray did not make use of his new Muslim name in Ottawa.

FAMILY BUSINESS

"Marrying Jamilah was Sean Murray's entry ticket to the Taib family business," said Ross Boyert. "Despite that, the Taibs have never really looked on Sean as part of the family. But there is no way out for Sean any more. He is imprisoned in the Taib empire and he is never going to be able to leave it again alive, even if he were to want to." Those are the words of an insider who was himself banished from the Taib empire—and who put up a fight against it, and suffered the full consequences.

Shortly after their wedding, Jamilah (27) and Sean (24) set up a jointly-owned company with its headquarters in Ottawa, in December 1987.[18] The company's objective was to manage real estate, in particular the buildings belonging to the "Sakto Development Corporation" ("Sakto"), which the Taibs had registered in Ottawa four years previously.[19] Back in 1983, just one year after finishing high school, Jamilah had become a director of Sakto, along with her younger brother, Abu Bekir, and her uncle Onn, who was one of Taib's brothers, the one whose job it was to pocket the kickbacks paid in Hong Kong for the export of tropical timber from Sarawak (see chapter 5).

The Taibs' real estate portfolio was quite extensive. In the course of its first year of trading, Sakto purchased more than 400 apartments in and around Ottawa.[20] In the following year, the Taibs acquired a large plot of land in the western part of Ottawa's city centre and set about planning an all-glass office tower, 46 metres high and costing 15 million dollars. The building was completed in 1989.[21] Sakto's financial statements show that the company had significant losses virtually every year for the first decade of its existence. Despite that, the value of Sakto's real estate quickly grew to more than 10 million US dollars in only a few years.[22] Not bad for equity capital of only 10,000 dollars.

In the early 1990s, once his marriage to Jamilah had stood the test of the first few years, Sean became a Sakto director. That was where the real money lay. It was, however, still Taib's brother Onn—or Taib himself in the background—who had the real say in things. By 1992, the value of Sakto's real estate had grown to 40 million Canadian dollars, against debt totalling 48 million.[23] These loans came from the Taib family itself, as well as from its secret financial companies in Hong Kong and Jersey. This is called a "loan-back scheme", one of the classic methods of money laundering: A criminal entrepreneur simply borrows his own money from shell companies controlled by himself.[24] Later on, Sakto contracted additional mortgages for 73 million dollars from Manulife, a large Canadian insurance company.[25]

One of my central arguments in this book is that while Taib and his family have committed untold crimes in Sarawak and under Malaysian law, they have also committed crimes in the process of moving their fortune abroad, including, but not limited to: fraud, tax evasion, and money laundering. Who the holders of Sakto's share capital are has never been disclosed until now. Canada's lax legislation makes it possible for a company's shareholders to remain anonymous. In 1989, Jamilah said they were a group of investors from Australia, Hong Kong and Malaysia who were "looking for a secure, long-term investment for their money and someone they trust to handle it for them".[26] There was, however, no doubt that, from the very beginning, Sakto (like Sakti in California) was a Taib family undertaking, and, if we believe Ross Boyert, it was personally controlled by Taib himself. In Canada as in the USA, the company had great success in attracting government tenants; the Canadian federal government itself, and

TAIB'S SECRET REAL ESTATE EMPIRE

2
SAKTI INTERNATIONAL GROUP
SAN FRANCISCO

Sulaiman Taib
Sean Murray

3
WALLYSONS INC.
SEATTLE

Sulaiman Taib
Sean Murray

W.A. Boylston
W.A. Everett
Sakti International
Corporation

4
SAKTO GROUP
OTTAWA

Jamilah Taib
Sean Murray

Adelaide Ottawa Corporation
Preston Building Holding Corporation
Sakto Corporation
Sakto Development Corporation
Tower One Holding Corporation
Tower Two Holding Corporation
Urban Sky Investments
Urban Sky Europe

1

TAIB MALAYSIA

Godfather of Taib family holdings

5

CITY GATE INTERNATIONAL CORPORATION OTTAWA

Jamilah Taib
Sean Murray
Thady Murray

100 %

6

RIDGEFORD PROPERTIES LONDON

Chris Murray

7

SITEHOST PTY ADELAIDE

Jamilah Taib
Sean Murray

N

no fewer than eleven of the ministries of the Ontario provincial government have leased space in Taib buildings.[27] At the time, this seemed the best possible evidence that everything about Sakto was completely in order.

Jamilah's connection with Sean—the son of a respected family of architects—was the ideal camouflage for the Taibs. It was the perfect opportunity to step up Sakto's business. Many believed that Sakto's properties were actually owned by the well-known architects of Murray & Murray. Further weight was given to that impression by the fact that gradually more and more Murray brothers and cousins took on jobs at Sakto. Sean's elder brother, Thady, took over the presidency of Sakto's sister company, City Gate International. His cousin, Chris, was catapulted to the top of Ridgeford Properties in London. And another cousin, Brian, took charge of the rental side of the Sakto properties in Ottawa.

The truth is that the famous architects—Sean's father Tim, and his uncle Pat—never had any role in Sakto, and there is no indication that they ever invested money in Sakto either. Many years later, they sold Murray & Murray to a large international architectural group.[28]

The Sakto group still appears in public today—along with its "sister companies" City Gate International in Canada, Ridgeford Properties in London, and Sakti in the USA—as a family business owned by the Murrays. Insiders like Ross Boyert know, however, that it is just a smokescreen, which the Murrays have helped create in order to avert critical eyes from the true ownership, as well as Taib's secret control over a global real estate empire worth hundreds of millions of dollars.

RESTRUCTURING

We return to Ross Boyert's account: "When I took over at Sakti at the end of 1994, Sean Murray and his brother-in-law, an architect from Ottawa, were working on a contract to renovate Sakti's head office in San Francisco. However, the project ran out of control financially, and I had to put a stop to it only a few months after taking on my new job.[29] Sean hated me for doing that, and I had no contact of any sort with him for the next ten years."

Ross's contacts with Sulaiman also became few and far between. "It was extremely difficult to get hold of Sulaiman on the telephone at that time. He kept saying that he was travelling to China and other countries, and whenever I finally managed to get through to him, our conversations lasted half a minute at most."

In those days, Ross was left more or less to his own devices in managing Sakti, and it was only occasionally that he reported to Sulaiman, whom Taib had entrusted with new, important tasks in Malaysia. Sulaiman was Taib's favourite child and had been hand-picked to become his successor one day.

In 1982, Sulaiman had followed his older sister, Jamilah, and older brother, Abu Bekir, to Ottawa, where he also attended Ashbury College and completed his secondary education there in 1986. Grandfather Abu-Bekir Chalecki, Taib's father-in-law of Lithuanian origin, had also taken up residence in Ottawa. The retired doctor had made the move from Adelaide, Australia, to Canada, where he was able to keep an eye on the grandchildren.[30] No wonder that the freedom-loving Sulaiman soon seized an opportunity to study in California, at the other end of North America, where it was much easier for him to escape from his grandfather's watchful eye.

After his playboy years as a student in San Francisco—Sulaiman graduated with a Bachelor's degree in business administration—Taib called his son back to Sarawak in the mid-1990s. Sulaiman and his young family returned home early in 1995, and he was immediately given a directorship in a family business, Cahya Mata Sarawak (CMS).[31]

Four years after his return to Malaysia, Sulaiman moved into banking and became a board member of the family-owned Utama Banking Group (UBG). In May 2003, by which time he had risen to president of CMS, he also took over the chair of the RHB Bank, Malaysia's fourth largest, which the Taibs had bought from the Malaysian businessman Rashid Hussain for 1.8 billion ringgits (roughly 540 million US$)[32] two years earlier.[33] Thanks to Taib's good connections with the Malaysian royal family, Sulaiman had also had the title of "Dato' Sri" ("Sir") bestowed on him, an honour akin to knighthood.

The turning point came in 2005, when the career of Taib's son, then aged 37, suddenly suffered a significant setback. Malaysia's central bank refused to confirm Sulaiman as chairman of the RHB Bank. The central bank, Bank Negara Malaysia, which had gained a strong and independent supervisory authority in the aftermath of the Asian financial crisis of 1997, obviously regarded Sulaiman as unsuitable for a position of such responsibility at the head of one of the country's biggest banks.

For Ross Boyert, who had been appointed and protected by Sulaiman, the glitch in the career of Taib's favourite son had calamitous consequences. Taib lost his trust in Sulaiman's business acumen and ordered a reorganisation of his real estate portfolio abroad. In September 2005, Sean Murray, operating from his base in Canada, took over responsibility for the Taib properties in the USA. Sean Murray had not forgotten the conflicts with Ross Boyert ten years before. He was eager to settle accounts.

"Taib in person instructed Sean Murray to take over the management of my firms," Ross relates. "At first, I simply refused to believe that I could be dropped so easily, but Sulaiman finally sent an official notice of dismissal. It was a gigantic shock for me. I was so angry that I referred to Sulaiman as 'Judas'."

Ross was worried about what would happen to his share of the Seattle profits, which Sulaiman had only promised verbally. Ross resisted handing the business over to Murray—but in vain. At the end of October 2006, the Taibs' lawyers handed him a document containing the signatures of all the Sakti shareholders, in which they declared their approval of the nomination of Sean Murray as sole director.[34]

In a last-ditch attempt to get things straightened out, Ross Boyert sent a long letter to the chief minister in person at the end of 2006, to which he appended two hundred pages of documentary evidence of all he had done for Sakti. Taib did not, however, respond, and Ross's resistance was pointless. He was dismissed. Two months later, he filed a suit with the Superior Court of California in San Francisco against his former employer, Sakti, and the Taib family on grounds of breach of contract, fraud, and infringement of labour law. With his statement, he included a detailed description of the properties owned in the USA by the Taibs, and the disguising of that ownership through offshore companies in the Caribbean and the Channel Islands.[35]

"That was the beginning of martyrdom for us," said Ross, looking back three years later. "After we filed the suit, our house was broken into several times, our car was vandalised, we received death threats, and we were constantly shadowed. But I regret nothing. I simply had to file that complaint for the sake of justice, and I'm determined to fight against Taib to the very end."

Boyert clenched his fists and finished by quoting from a classical Greek author: "What was it that Leonidas of Sparta replied when the Persian king, Xerxes, called on him to lay down his arms rather than suffer defeat in his last forlorn battle? *Molon labe*! Come and take them! I am not going to give up!"

One thing became clear in the course of our conversation with Ross and Rita Boyert: The two of them were beginning to break down from the constant pressure, and were in urgent need of support and recuperation. But were Clare and I, I wondered, really in a position to assist them? I had a premonition that help might have come too late for Ross.

PERSECUTION COMPLEX

By then it was midday in California, and despite the electrifying report from Ross and Rita Boyert about their activity for the Taib family, we were urgently in need of a break. Clare switched off her microphone and we left the basement meeting room in the Marriott LAX.

Our first thought was to secure the documents proving Taib's ownership of the Sakti real estate group. While Clare looked after the Boyerts, I took the pile of documents up to the hotel's business centre on the ground floor. I made one set of photocopies for the Bruno Manser Fund, one set of copies for Clare, plus two sets of scans on CDs. I packed the CDs in envelopes, took them to the FedEx counter in the hotel, and sent them off to London and Switzerland. It was a precaution against theft or lost luggage. The multiple copies would ensure that we would be able to hold on to this valuable evidence.

While I was still making the copies, it struck me that a thin man was prowling around in the hotel lobby. With a look of apparent disinterest, he walked into the business centre, and eyed me and the documents with an uneasy expression on his face. He did not utter a single word, and, since I was too busy checking through the copies, I did not bother any more about him. Before long, he disappeared.

Clare and the Boyerts came along to the business centre, and Ross and Rita were visibly agitated. They had seen the same man, and they regarded him as a Taib spy. "There's a GPS in our iPhone," Rita said. "That's probably how he found us."

Why on earth didn't she simply switch the phone off? was my first thought. "Look," continued Rita, as if she had heard my unuttered question. "I can't turn off my telephone." Rita Boyert pressed the off button on her iPhone. The display went dark. Less than twenty seconds later, however, the device lit up again all by itself, as if it had never been turned off.

On one of the innumerable occasions on which the Boyerts' home had been broken into, the iPhone had disappeared, but later it had suddenly reappeared. Rita and Ross suspected that while it was gone the settings had been copied, and the phone replaced with a duplicate. Whoever was following them was able to eavesdrop on their calls, locate them at any time, and manipulate the device through remote control. "All our electronic devices are being watched," Ross said. "We got used to that a long time ago. Since we want our daughter to be able to reach us at any time, we have to accept compromises. Even a new phone would only improve our privacy for a short period of time."

One thing that Ross did not mention was that he did not have enough money for a new mobile phone. That became clear later on, when we were leaving the hotel car park in the Boyerts' black Jaguar, the only status symbol still left to the couple who used to be well-off. When we stopped at the barrier to pay to leave the car park, Boyert asked if I could give him a few dollars. His wallet was empty. The man was bankrupt, and yet he was still driving a Jaguar.

We agreed that we would all fly together to San Francisco that same day, since the Boyerts had collected further documents there that they wanted to show us. First of all, however, Ross took us to Costa Mesa, a Los Angeles suburb, where the couple had found temporary accommodation after being evicted from their house. There was only time to pack their suitcases with the barest of essentials and to collect their dog, a long-haired pinscher called Scipio. After that, we immediately continued to Los Angeles airport.

There were only a few passengers on board the Virgin America afternoon flight 933 from Los Angeles to San Francisco, and we sat near the back of the plane.

The four of us browsed through pictures of properties owned by Taib in Canada, England, and Australia. Clare had found them on the Internet and printed them out. There was the eight-floor building "The Adelaide" with 158 luxury apartments in Ottawa; three huge office towers in a part of the Canadian capital known as "Little Italy"; a hospital in Fitzrovia, an upmarket district of London; the massive limestone palace at Tokenhouse Yard next to the Bank of England; the Hilton Hotel in Adelaide, Australia; and many more.

As we talked with Ross Boyert, Clare and I both felt a growing concern about the health of the former Taib employee. Suddenly, he expressed serious doubt whether it had indeed been he who had negotiated the rental contract with the FBI for the big Taib building in Seattle. He asked: "Was I really the decisive person in that deal with the FBI or did the negotiations actually take place at a higher level without me being informed? It is feasible that the US government made a secret deal with Taib. Who is really behind the FBI, the CIA and the US government? Are they secretly under Taib's control?"

I had not imagined Ross's persecution complex to be so serious. Yet the man had been constantly intimidated for several years. Real and imaginary situations were becoming mixed up in his mind. And, least of all, I could not imagine that the fear of being followed could be so strong here in San Francisco.

"That's him!" Rita Boyert hissed and pointed at a tall, bald-headed man near the exit. "That man's been stalking us for years. He even pursued us on a vacation to Hawaii, when we tried to get away from all of this. He just sat there at a table in our resort with a broad grin on his face, knowing full well that our holidays had been ruined." Earlier that morning, Rita and Ross had already shown us pictures of the heavily built man. It was eerie to see him now right in front of us.

In a taxi, we set off to find somewhere to spend the night. We had not travelled far when we noticed a grey car following us at some distance. Even after we turned several corners it was still there. At last we arrived at the Embassy Suites, but found it already fully booked. The kind woman at the reception recommended another hotel nearby belonging to the same chain. Before setting off again, however, Rita Boyert deposited her iPhone at the reception in an attempt to give whoever was following us the slip.

Our taxi was still waiting in front of the hotel, and once the luggage had been stowed again I suddenly saw a man looking out of the hotel lobby in our direction. My reaction was to pull out my camera (the only weapon I had against stalkers, real or imaginary) and to press the shutter. Once, twice, three times. He disappeared instantly when he realised my camera was pointing at him. But now I had proof that he had not been a hallucination, but flesh and blood. And someone had paid that man and given him instructions to stand there and keep an eye on us.

SKUNK ATTACK

"We are talking about an extensive and very costly operation," said Ross Boyert, with his vibrant voice. "It's designed to break you down. It's intended to get you to the point where you're so discredited that no one is going to believe you. We filed a suit against Taib, supported by ample evidence. But they are doing their best to discredit us and keep this case away from the public." Ross's paranoia showed how close to its goal the campaign against the whistle-blower had already come.

Evening had fallen in San Francisco, and we were having dinner in the Boyerts' hotel room. We had turned the television volume up so that no one could eavesdrop on our conversation. It was a ploy that people in East

Germany had used against the Stasi, and we too were using it—here in freedom-loving California against the long arm of a corrupt ruler far away in Borneo.

The time had come for the Boyerts to tell us in detail how they had been intimidated and terrorised for three years, and the details appalled us. Twice hubcaps had been hurled at their car while they were driving along the highway. Once their windscreen had been smashed, and the other time another car had been hit. "I think they were taking a great deal of pleasure in what they were doing to us," said Ross. Their house had been broken into several times, yet nothing had been stolen. Only the pictures on the wall had been switched around. How can you go to the police with a story like that?

There was a long list of incidents. The Boyerts' daughter had been served a spiked drink in a bar that had knocked her out. Rita Boyert had received a telephone call telling her that all she needed to do was to divorce Ross and all the harassment would immediately stop. Rumours of alleged wrongdoings by the Boyerts, and of related police investigations were spread among business partners and acquaintances. The Boyerts found their friends gradually turning their backs, no longer wanting to meet with them. After all, who would want to have anything to do with people who seemed to attract so many problems? In the end they found themselves totally isolated and alone.

The Boyerts made more and more complaints at the police station in Atherton, where they lived, but they did not receive any serious support from the authorities. Of the many politicians and decision-makers whom the Boyerts asked for help, only the Jewish Congressman and Holocaust survivor, Tom Lantos, really took them seriously. However, when Lantos died soon after at the age of eighty, in February 2008, the Boyerts lost that hope too.

After hearing the Boyerts' disturbing story all in the course of one day, I had difficulty getting to sleep that night. But I finally managed it, with the Sakti documents well-packed in my suitcase and my alarm clock set. Then all of a sudden in the middle of the night, I woke up with a shock, instantly wide awake. I had smelled gas in my room. I quickly switched on the light, alerted the reception and hurried out of the room in my pyjamas. Was that a warning from the Taib thugs to keep my nose out of their business, or was my imagination beginning to play tricks on me too?

"It was probably a skunk," said the gruff-voiced hotel employee, who came up to my room, opened the window to let fresh air in and handed me the key for another room.

Can a skunk's stink penetrate closed windows on the sixth floor of a hotel, I wondered. I decided to spend the rest of the night in the hotel lobby. I got dressed quickly and packed my suitcase. It was only two hours until dawn, and the combination of jetlag and coffee kept me awake. I had rarely been so happy to see the sun rise, as the hotel guests emerged from their rooms to converge on the breakfast buffet.

GRIEF AND TEARS

Every seat was taken in the Swiss Airbus bound for Zurich as it taxied to the runway. I breathed more easily as the acceleration pressed me back into my seat, and the aeroplane took off into the night sky over North America. I had rarely been so relieved to leave a place. I had abandoned my original plan of extending my stay for a few days' holiday in the USA. My only thought was to get away.

I had said goodbye to Clare and the Boyerts an hour earlier, and they had set out on their return journeys to London and Los Angeles respectively. In San Francisco, the Boyerts had shown us all the Taib properties and also their former house in Atherton, which stood abandoned with the unmowed lawn growing wild. We then went to Menlo Park, where Ross kept duplicates of all Sakti company documents in a rented storage room. We copied the most important documents and would soon publish them on the Internet.

Ross Boyert's twelve years of loyal service to the Taibs were now at an end—and he had paid a high price for it. "Active involvement in the milieu of organised crime always ends in tears, grief, death, prison, blood, and poverty,"[36] said Pietro Grasso, head of the special anti-Mafia prosecution bureau in Italy. Something similar could also be said of Taib's global network of environmental destruction, corruption, and money-laundering. Ross's role in Taib's criminal network was little more than that of an extra, but nonetheless significant enough to trigger a harsh vendetta. Moreover, he must have known from his very first day at Sakti that he was playing a dangerous game.

"If they had killed us, it would have been a blessing rather than driving us slowly insane in this way," said Rita Boyert, at the height of her despair. For Taib, however, crude murder would have been too risky. He had chosen to terrorise them psychologically, to isolate them from their friends and to destroy their means of subsistence. It was, however, unclear who was implementing this on Taib's behalf in the US.

After the visit to the Boyerts in California, Clare and I spoke with them often. A generous donor was found to help them out of their financial straits with a few thousand dollars. A security company was hired in California to safeguard the couple if they felt threatened. Ross, however, suspected that the security personnel might turn out to be a new threat, and, in the end, he dismissed them.

We drew up a plan to bring the Boyerts over to Europe for a holiday to recuperate, and to establish contacts with opposition figures from Malaysia. But at the decisive moment, Ross refused to board the aircraft and to leave the USA. He had suddenly lost his trust in Clare Rewcastle, and began to suspect that—because of her prominent brother-in-law—the British government or some other dark agency might be behind her. Ross was trapped in his paranoia, and could no longer find a way out.

Two months after our visit, Rita Boyert phoned Clare Rewcastle with an urgent message. Ross Boyert had wrecked his Jaguar in a suicide attempt and had been admitted to a psychiatric hospital. That same day, Clare flew to San Francisco to do what she could for the depressed former Taib employee. All help, however, was in vain. A few weeks later, on Sunday, 3 October 2010, Taib got what he wanted. Ross Boyert was found dead in a hotel in Los Angeles. Ross had pulled a plastic bag over his head, taped it tight around his neck and suffocated. He had killed himself.

PARADISE LOST

Since times immemorial, the Penan have been roaming the rainforest of Borneo as nomadic hunter-gatherers. Generations of researchers have been fascinated by these forest dwellers and their unique environment. And then loggers arrived on the scene, and began to destroy the last paradise on Earth.

A HEADMAN REMEMBERS

Heavy drops of rain thundered on the metal roofs of the huts in Long Gita as night fell in the rainforest of Borneo. The chorus of the cicadas in the rainforest's giant trees had already come to an end, and now it was possible to hear the croaking of the tree frogs in the distance. The black night was setting in on one of the loneliest corners of Sarawak.

The face of Along Sega stood out in the light of the glowing fire. The impression made by the Penan headman was one of fatigue and sadness. A strenuous, eventful life had left its traces. Along's headdress, made of the black-and-white feathers of the rhinoceros bird, hung from the beam holding up the hut, while his ear ornaments made of wooden discs lay on a rattan mat woven by his wife from the fibres of forest plants. The agile hunter's blowpipe and gun leant against the rough timber-planked wall.

Along stared pensively into the flames. His expression didn't change as he recalled his childhood in the then-pristine forest. "My birth wasn't registered anywhere. All I know is that I am now already older than seventy." He was born in the rainforest at some time in the late 1930s, when no one in this forest even spoke of years or hours. Traditionally, the forest-dwelling Penan have been guided solely by nature's annual cycle, with the rainy season and the dry season, plus the fruiting season, which happens once every few years when the rainforest trees all blossom in spectacular abundance at the same time and then become heavy with fruit, as if reacting to some hidden signal.

Along's birthplace was not far from Long Gita, where his kinship group was based at that time in the upper reaches of the Limbang River. In the indigenous language, "long" is the word for river mouth, and, as the Penan often settle near river mouths, many village names contain the word "long" or, alternatively, "ba"—the term for water.

As the youngest of six siblings, Along was still just a child when his father, Sega, introduced him to hunting. He had a broken-off branch as his first spear and a bamboo tube as his first blowpipe for target practice. It did not take long for the young Along to acquire sufficient skill to be allowed to hunt for animals by himself.

"When I got better at aiming the blowpipe, my father took me to hunt birds near our encampment. Later on, he gave me a spear for hunting wild boar—and then showed me how to process the pith from the sago palm. While I was still very young, my siblings and I learned everything about how to survive in the forest."[1]

As the last nomadic hunter-gatherers of Southeast Asia, the Penan know the Sarawak rainforest like the back of their hands. In groups of sometimes fifty, they have lived in the rainforest and roamed the mountainous heart of Borneo for centuries. They have explored hundreds of rivers, mountain crests, and hunting trails through the rainforest, and given them all names in their own language. One hectare of forest here is home to more tree species than all of Europe, and the Penan recognise more than a thousand species of plants, and have given them names, and learnt how to make use of them.

The most important plant in the Penan culture is the sago palm, *uvut*[2], which grows wild in the rainforest and whose pith contains large amounts of pure starch. The Penan use sago flour to cook *na'o*, a viscous, almost tasteless porridge, which they sometimes flavour with the fat of wild boar. Why go to all the bother of cultivating rice when they already have sago as a gift of nature? The second basic plant is *tajem*[3], the dart-poison tree, from which the Penan obtain the lethal poison for hunting with the blowpipe. It acts so quickly that a monkey hit by a poison dart is dead in a matter of minutes. Once the poison gets into the bloodstream of a mammal or human being, there is no way of saving them apart from the juice of an inconspicuous herb and natural antidote, which the Penan call *getimang*.[4] Centuries ago, the original inhabitants of Borneo discovered its effect and have handed down that knowledge from generation to generation ever since.[5]

As a young man, Along used to attend the *tamu* market organised every few months by the British colonial administration. The Penan would exchange their rainforest products for other goods from neighbouring peoples and traders from the town. "We used to sell rattan fibres for making mattresses, and the curved beaks of rhinoceros birds as jewellery," Along recalled. There was a particular demand for tree resins, and bezoars—stones monkeys have swallowed to aid digestion, and which were an important ingredient in Chinese medicine. "We used to take resins from the forest for the Murud, a neighbouring people from the same region, and in exchange we got dogs from them, which we used for hunting."

The Penan exchanged these items for cooking utensils, tools, and guns. The *tamu* markets used to take place every four months. "Since the Penan didn't have a calendar in those days, we used a rattan string, in which we tied 120 knots, untying one every day. When there were only a few knots left, we used to pack our things together and set off. That made sure that we were always in time for the *tamu*."

"We have been living in the region around the Batu Lawi mountain for a very long time," Along added, peering into the darkness, upriver towards the mountains. There in the distance, the two limestone columns of Batu Lawi rose into the night sky. At that point, he lowered his voice and continued quietly—something the Penan always do when they have important things to say. "Batu Lawi is the home of the good spirit, Shinan, who lives in the rock over there. That is how the two Batu Lawi peaks came into being. A long time ago, the two rocks on top of Batu Lawi used to be two people, a man and a woman, who were very much in love with one another. But the evil spirits became jealous of them and killed first the woman and then the man. They then both turned into solid rock. My grandfather, Tawin, and my grandmother, Bresen, were the guardians of the Batu Lawi region. When they died, my father assumed the role. Today, I am responsible for it."

The ancestors of today's Penan used to revere not merely Batu Lawi, but other mountains, such as the 2,423-metre-high Gunung Murud. They also cherished certain fruit trees or regions in the rainforest that are particularly rich in plant and animal life. The Penan use the term *molong* for their tradition of claiming particular regions, trees or plants for their kinship group: "If the first person to see a fruit tree in the rainforest marks it with his *parang*, his machete, then everyone else knows that the tree is under that person's claim. Anyone wanting to eat its fruit will first of all have to ask permission," Along explained.

The Penan not only make use of trees, forests and mountains for their daily needs, they have a strong spiritual attachment to them as well. "We have many secrets in the forest where our ancestors are buried."

Before their conversion to Christianity, the Penan used to believe in numerous spirits, most notably the eagle. If the eagle signalled with its wings, it was likely a bad omen.[6] In order to appease the eagle-spirit, the Penan used to carve wooden *seperuts* or charms for him and to place them in the forest. Along suddenly broke off his account. He preferred to talk about people rather than the threatening world of the spirits.

The first strangers Along ever met, when he was an adolescent, were members of neighbouring indigenous peoples, who lived from cultivating rice. During the Second World War, he once saw a Japanese soldier from a distance in a jungle settlement. The first white man he ever met was John Griffin from the United Kingdom, who for more than twenty years was responsible for the Penan territory in the White Rajahs' administration of Sarawak. Later, Australian missionaries from the Borneo Evangelical Mission came to the region.

"We were very well off at that time. The forest was still intact, and we lived off its bounty. Everywhere there were sufficient numbers of wild boar, lots of sago palms and plenty of clean water for preparing sago. The forest was a paradise for us nomads in those days. If we came across a place that we liked the look of, we would stay for several months, and live off the ripe sago palms in the surroundings." Once most of the sago palms had been harvested, the Penan would pack their handful of possessions and move on, walking for one or two days to a new location.

That, however, was back in the days before Taib became chief minister of Sarawak, and before the Limbang Trading Company began to clear the forest in Along's territory. Most of the approximately ten thousand Penan had already settled into villages by that time, following the urging of the missionaries and the government. They had learned to cultivate rice. Only a few hundred Penan, like Along Sega, wanted to continue with the nomadic way of life.

Along was no longer a child when he met another foreigner, the 30-year-old Swiss visitor, Bruno Manser. An alpine shepherd searching for an alternative way of life without money, Manser made contact with the last Borneo nomads in 1984 and joined up with Along's group to get to know life in the rainforest and to document it in his diaries. The young Swiss and Along got on so well that Manser became Along's adopted son. The fascination of living in nature bound the two friends together, although their backgrounds could hardly have been more different. Later, they fought side-by-side against the destruction of their earthly paradise.

BORNEO'S RAINFOREST

Of all the world's forests, Borneo's tropical rainforest is one of the oldest and most beautiful. The different species living there are more numerous than almost anywhere else in the world. Whoever has ever set foot in this virgin forest will feel the urge to return. The crowns of the giant trees belonging to the dipterocarpacean family grow so closely together that the illumination of the forest floor resembles permanent twilight, where the voices of the forest animals can be distinctly heard: the courtship call of wild peacocks, the raucous croaking of rhinoceros hornbills, and the guttural call of startled macaques. Tens of thousands of insect species, hundreds of bird species, and dozens of different mammals inhabit this archaic habitat. It is one of six world regions with the highest biodiversity. Orangutans, Malay bears, tree leopards, and many other shy creatures inhabit the dense jungle of inland Borneo. Carnivorous pitcher plants, orchids, wild bananas, and colourfully blossoming ginger plants flourish on the tops of trees and in the undergrowth. The Borneo forest has existed since time immemorial, and human beings have been living there for at least forty thousand years.

In 1869, describing the Borneo rainforest, the British naturalist and explorer Alfred Russel Wallace (1823–1913) noted in his travel journal *The Malay Archipelago*: "For hundreds of miles in every direction, a magnificent forest extended over plain and mountain, rock and morass."[7] He continued: "The forests abound with gigantic trees with cylindrical, buttressed, or furrowed stems, while occasionally the traveller comes upon a wonderful fig-tree, whose trunk is itself a forest of stems and aerial roots."[8] While travelling along a river, Wallace saw "the beautiful virgin forest coming down to the water's edge, with its palms and creepers, its noble trees, its ferns and epiphytes".[9]

The Welshman Wallace, a contemporary of Charles Darwin, thoroughly explored the islands of Southeast Asia in the middle of the 19th century, covering a distance of some two thousand kilometres. A photograph taken with his mother and sister before he left London shows the full-bearded 30-year-old in 1853, wearing evening dress, his thick dark hair parted on one side, and gazing intensely through the thick lenses of his pince-nez. Wallace had recently returned from a four-year expedition to the Amazon. Now he was eager to get on with exploring the jungles of Southeast Asia.

On 1 November 1854, Wallace landed at the mouth of the mighty Sarawak River in the north of Borneo, where he was welcomed by James Brooke, the British "White Rajah" of Sarawak (see chapter 3), who kindly invited Wallace to stay with him. Wallace was so enthused by Sarawak that he remained there for more than a year, and made many expeditions into the virtually impregnable interior of the country. He carried out intensive zoological and botanical studies. He tracked down orang-utans and shot some of them as specimens for his collection. He paid rewards to locals who brought him beetles and tropical butterflies. He was so taken with durian, a tropical fruit growing wild in Sarawak and known, in particular, for its incredibly strong odour, that he recommended his readers to travel to the Far East solely to experience its very special flavour.[10]

Alfred Wallace returned to Britain in 1862, after eight years in Southeast Asia, bringing with him some 125,000 natural-history specimens. His collection of beetles alone numbered more than 80,000. Science's debt to Wallace's expedition includes the discovery of more than a thousand new animal and plant species that he was the first to describe. One of the most peculiar of these is a gliding tree frog known as Wallace's flying frog[11], which lives high up in the rainforest trees and, thanks to membranes between its toes, is able to glide up to 20 metres through the air.

It was through his critical analysis of the multitude of flora and fauna on Borneo that Wallace developed theories on the origin of the diversity of species. These were astonishingly similar to those of Charles Darwin, though Wallace arrived at them independently. During his stay in Sarawak, he wrote an article in 1855, in which he formulated what was later called the "Sarawak Law" of evolution, according to which "every species has come into existence coincident both in space and time with a pre-existing closely allied species."[12] After another three years' research in Southeast Asia, Wallace came to the conclusion that natural selection had to be the explanation for this evolution and he adressed a paper on the subject to a then-relatively unknown scientist in Britain, named Charles Darwin, who was fourteen years his elder.

Twenty years previously, Darwin had undertaken his legendary expedition to the southern hemisphere on board the *Beagle*. It was the evaluation of his extensive research material from the Galapagos that had led him to the pioneering conclusion that a law of nature was the cause of biodiversity, and not a divine creator as recounted in the Bible. Darwin described this law of nature as evolution through mutation and natural selection. But for very many years, he had kept this scandalous theory to himself, and only now that Wallace had arrived at similar results independently, did Darwin realize that he would have to hurry his publication.

At Darwin's instigation, Wallace's paper was read out to a meeting of the Lynné Society in London on 1 July 1858 along with an excerpt from Darwin's unpublished manuscript on natural selection. In November 1859 (while Wallace was still away in the Far East), Charles Darwin's epoch-making work *On the Origin of Species* was published, setting out his theory of evolution.

While Charles Darwin's name is writ large on the pages of history, that of Alfred Wallace, the natural scientist who travelled to Sarawak, has largely been forgotten. Bio-geographers are familiar with the so-called Wallace Line, defining the boundary between Asian fauna and the very specific Australasian fauna. In one of his other works, Wallace also went on record as a visionary: His *Tropical Nature and Other Essays*, published in 1878, warned against felling the tropical rainforest and the resulting erosion of the soil.[13]

NOBLE SAVAGES

Three decades after Wallace, another young man with a passion for natural history travelled to Sarawak from Britain. He was Charles Hose (1863–1929), destined to make a career as a civil servant working for the second White Rajah of Sarawak, Charles Brooke. Hose is commemorated by the mountain ridge in Central Sarawak bearing his name. Likewise, a dozen vertebrates, most of which are native to Borneo, are named after him. These include frogs, a monkey, a viverrid, and two species of birds (Hose's broadbill and Hose's oriole). Most of these animals are today on the Red List of species threatened with extinction as a result of the loss of their habitat through rainforest destruction.

Hose, the son of an English country clergyman, had just begun his studies at Cambridge. He was finding little enthusiasm for it when his uncle, the Anglican Bishop of Singapore, negotiated a position for him as an administrative cadet in the Sarawak colonial civil service. The 21-year-old did not hesitate: Abandoning his studies, he boarded a ship bound for Borneo. What Hose missed in Cambridge was more than made up for by what he taught himself in the Sarawak rainforest. He turned out to be a fastidious observer and disciplined zoologist. After nine years in Sarawak, he published a book on the mammals of Borneo, which remains a standard work of reference.[14]

After studying the animal kingdom, Hose turned his interest to human beings and compiled an extensive anthropological report on the indigenous inhabitants of Borneo. Later, he expanded this into a two-volume book entitled *The Pagan Tribes of Borneo*.[15] Hose stayed in Sarawak for nearly a quarter of a century, of which sixteen years as the senior district officer in Claudetown (today's Marudi) on the Baram River, the gateway to one of the island's least-known and least-explored regions.

At the turn of the 20th century, Borneo was regarded as an island swathed in myths, on the very edge of civilisation, with headhunters on the rampage in the dense jungle—one of the last remaining empty spaces on the maps of the European colonisers. Novelists like Joseph Conrad, H. G. Wells, and later Somerset Maugham placed their characters here, at the edge of civilisation. Through his studies and publications about the indigenous peoples of Sarawak, Charles Hose made a major contribution to dispelling western prejudices about the "savages" in Borneo. At the same time, however, he introduced new and potent *idées fixes*.

Hose was the first to describe in detail the life of the people in the jungle and on the rivers of Sarawak, particularly those on the Baram, a mighty river system that has its source in the jungle, at the heart of Borneo, on the present-day border with Indonesia. The indigenous Iban, Kayan, Kenyah, Berawan, and Kelabit peoples lived here. Hose visited their longhouses (large communal buildings, in which each section was occupied by a different family), studied their social system, their farming, their tools, their handicrafts, their jewellery, and their tattoos and got them to tell him about their wars and conflicts. One of the cultural peculiarities was that many longhouse occupants used to go on ritual headhunts against other indigenous peoples. The victims' skulls were hung in prominent locations inside the house as a symbol of social prestige.

He firmly believed that an enlightened colonial master should understand his subjects. Despite the fact that he had no academic qualifications, his outstanding scientific talent made him in only a few years an internationally respected authority in Borneo research.

Hose was particularly enthralled by a remote nomadic forest people in the upper reaches of Sarawak's rivers, that no European had described before and that he called "Punan". He suspected that they were the genuine aboriginal population of Borneo.[16] Describing these shy forest people, Hose wrote: "The skin of a Punan is of a fine silky texture, and is either of a pale yellow or even of a greenish shade; for the Punan rarely exposes himself to even indirect sunlight, preferring always the deep twilight of the jungle."[17]

"They are great hunters, being able to move through the jungle without making the slightest noise, and have a name for every living thing, known even by the small boys. They are wonderfully expert in the use of the blowpipe, shooting their poisoned arrows with such precision that it may be said that they seldom miss even the smallest object they aim at, yet this efficiency with their weapons notwithstanding, they are a very timid race, but will fight in self-defence."[18]

The "Punan" were the only people living in Sarawak who were not warriors and headhunters, who did not engage in farming, did not live in longhouses, kept no livestock, and did not use boats. They lived solely from hunting with the blowpipe, from processing sago, and from gathering forest products. "When to these curious facts it is added that they in every way come up to the ideal of the 'gentle' or 'noble' savage [...] surprise increases." Charles Hose described his "noble savages", in whose facial features and posture he also saw "something of the air of an untameable wild animal",[19] marked not only by great simplicity but also high moral standards: "They are an honest and unselfish people [...] and when once well-known they undoubtedly prove to be the best-mannered people of any of the savage tribes inhabiting the island."[20]

Had Charles Hose found humankind in its unspoilt natural state deep in the interior of Borneo? In speaking of "noble savages", Hose chose a controversial term that had first been coined by the British playwright John Dryden in 1672 and that had been haunting cultural history ever since then.[21] The expression "noble savage" was often wrongly ascribed to Jean-Jacques Rousseau, who was known to believe that human beings are good by nature, that they had originally lived in the forest, and that they had only been morally undermined by the founding of private property and civilisation. In his culturally critical discourse on the *Origin and Basis of Inequality among Men* published in 1755, the Geneva philosopher does not use the expression "noble savage" at all, only "savage man" *(l'homme sauvage),* "the savage" *(le sauvage),* and "natural man" *(l'homme naturel).*[22]

The empirical evidence of "noble savages" in the jungles of Southeast Asia struck a chord for Europeans in the age of discoveries, and should not be underestimated as a feature of the marketing of Charles Hose's books. Towards the end of his life, Hose, who had returned to England, made the "Punan" more and more into a central element in his writings. *Natural Man* was the title he gave to the summary of his anthropological research in Borneo, published in 1926. In the preface by a professor friend of his, we read: "If his book achieves no other purpose than to establish the fact of fundamental importance that man is by nature peaceful and good-natured [Charles Hose] will have achieved a revolution in anthropological doctrine."[23]

However, contemporary science cared little for such a revolution. Many rejected Hose's revision of Rousseau's theory of natural goodness (*bonté naturelle*) of man. *Natural Man* nevertheless became a bestseller and still retains its relevance today.

A VEGETARIAN AMIDST HUNTERS

Professor Rodney Needham said: "Charles Hose was wrong when he called the Penan 'Punan', which he did because he had a Kayan woman as a lover. The Kayan use that name for the nomads living in the same region, but Penan with a *schwa* [Pə-nan] is what they always call themselves."

The old professor was leaning back on the couch in his living room. He spoke slowly and was short of breath, but, despite the effort involved, his speech was very clear. His facial expression betrayed concern. Every now and then he turned his head in my direction, and, with a severe look, his light blue eyes checked to make sure I was listening closely.

It was Tuesday, 26 April 2005, and one floor below him in the busy Holywell Street, Oxford's university life was bustling as usual. Couples were strolling arm-in-arm along the street, their bags weighed down with academic textbooks. Asian tourists were shuffling towards the Bodleian, Oxford's venerable university library. In the King's Arms on the street corner, students clutched pints of beer.

The Oxford University Expedition, sponsored by the British Museum, first set out to explore the Sarawak jungle in 1932.[24] Rodney Needham

(1923–2006) himself has spent more than half a century researching and lecturing on the peoples of the rainforest—with interruptions for field visits to Sarawak and Indonesia.

I asked the emeritus professor to tell me about his research into the Penan nomads. Eighty-two at the time, Needham was suffering from a lung disease, and had only a year and a half left to live. Every breath was a struggle. Despite that, he maintained his scientific curiosity and his willingness to help to the very end.

Needham is one of the leading figures of 20th-century British anthropology. His essays and books are still compulsory reading for students studying social anthropology. During the Second World War, Needham fought as a young captain for the British against the Japanese in Burma and only just survived the battle of Kohima, the "Stalingrad of the East", in 1944. After the war, he returned to Southeast Asia, this time on a peaceful mission, to research the Penan for his dissertation at Merton College.

Needham lived for several months with nomadic Penan groups in the central part of the Baram valley, and also learned their language. He kept painstaking notes of their names, the size of their kinship groups, their habits, relationships, and routes taken through the jungle. He wrote the results down in his spidery handwriting in dozens of field books, which formed the basis for his dissertation as well as several scientific articles. Needham's dissertation was never published, but it soon became the stuff of legend.

As a vegetarian, life was bound to be difficult for Needham with a hunting people, one of whose staple foods was the Bornean wild boar. His field stay with the Penan allegedly ended with a nervous breakdown. Nonetheless, he had many fond memories, and showed me a rattan bracelet set with glass beads, which he kept on his living room mantelpiece. A young Penan woman had woven it for him.

One of Needham's most important findings was that the Penan are divided into two very distinct groups, whose territories are separated by the Baram River. On the right bank (east) of the Baram, where the Selungo tributary flows into the Baram, is where the Eastern Penan *(Penan*

Selungo) live, while on the left bank (west)—the catchment of the Silat River—live the Western Penan *(Penan Silat)*.[25]

It is important to clarify that, in describing the Penan as mild-mannered, timid "noble savages", Charles Hose was referring solely to the Eastern Penan, who (like Along Sega) held out as nomads for the longest, and resisted deforestation with acts of civil disobedience. These are the people with whom my organisation, the Bruno Manser Fund, has been working since its creation, and it is generally they who are meant when the word "Penan" is used in this book. In contrast, the Western Penan wage their conflicts more openly and are known to be loud and outspoken. They have even been referred to as the "New Yorkers" of the rainforest.[26]

In the early 1950s, Needham estimated the Penan population to be as low as 2,650, two thirds of whom lived as nomadic hunter-gatherers. One third of them, nearly all Western Penan, had given up the nomadic way of life and become sedentary in villages, where they now lived "in a fashion similar to more familiar Bornean peoples".[27]

Even as an old man, Rodney Needham was still very much concerned about the welfare of the Penan. He consented to my request to file an affidavit with a notary public, in which he documented those areas of the rainforest in which he had encountered Penan nomads while doing his field research. This statement is intended to help the Penan in the struggle for their land rights, and for the general conservation of the rainforest.

On my second visit to Oxford, the professor gave me a very special gift: prints of black-and-white portrait photographs of Penan, which he had taken in the Sarawak jungle in 1951, the first-ever extensive series of photographs in the history of the rainforest nomads.

Not long after that, Professor Needham sent me a letter neatly written on his manual typewriter: "Ill health has prevented me from attending to [your last letter] but I shall do what I can when I am enough recovered. I am afraid that I have to say that a visit would be (as the previous two proved) too exhausting."[28] As things turned out, unfortunately, the eminent academic did not recover. News of his death reached my office at the end of 2006.

PENAN TRAGEDY

I had Rodney Needham's photographs and his affidavit in my luggage when I travelled to Sarawak in autumn 2006 to meet our Penan friends and their lawyers. After a six-hour journey along the logging road, I arrived at the Penan village of Long Bangan. I was hoping I would be able to find witnesses there to help me identify who was in Needham's pictures.

"No doubt that's me as a young woman," said Aren Umai, taking a very close look at the old photograph. The Penan woman beamed, but otherwise concealed her excitement. "Here's the proof. Look at the tattoo on the lower part of my left arm. Now look at the tattoo on the woman in the photograph. I got these five points tattooed on me, because five good things happened to me within a short span of time."

I had to admit she was right. The young, bare-breasted beauty in the black-and-white photograph, and the aging—but still spritely—Penan woman in front of me were identical. The date was 7 September 2006, an important day for Aren, and she was in a really good mood. The 70-year-old Aren had never before seen a picture of herself as a young woman and she was totally surprised. She couldn't even remember meeting Needham and posing for him. It is more than likely that at that time she would not have known what a camera was and how it worked.

The picture of Aren was taken fifty-five years before, in November 1951 at a *tamu* market in Long Melinau on the Tutoh River, a tributary of the Baram, which has its source high up in the mountains. "In those days we used to get decent prices for our goods from the forest," said Aren, and as proof opened her mouth to show several glistening gold teeth, which she had had a dentist put in for her when she was a young woman.

At the time of writing, Aren lived with her six children and numerous grandchildren only a few kilometres away from the UNESCO-protected Mulu National Park in the village of Long Bangan, which had a population of 400. The settlement in the lower reaches of the Tutoh River comprised sixty-eight simple houses built with dark timber planks and standing on stilts.

"Look at my dress in the picture, the *talun*," Aren said to the children, who had gathered around her, curious to see her old photograph.

"I made it myself from bark. That was typical of what we Penan Selungo women used to wear when we still lived as nomads in the jungle." As a young woman, Aren had been particularly proud of the large wooden discs in her earlobes, even though it was very painful to have her ears cut. Elongated ear lobes were regarded as a symbol of success.

In the meantime, the headman of Long Bangan, Unga Paren, had joined us. "In 1960, my father, who was headman at the time, founded this village, when he was looking for a place in which our Penan group could settle," he said. "We had to accept conversion to Christianity before settling, because our old belief in spirits would not have allowed us to spend a long period of time in any one place."

In the early days when Sarawak was governed by the "White Rajas", and later by the British Crown, the entire Penan territory was still covered in virgin forest. It was only after the withdrawal of the colonial power that deforestation began in the interior of the country. "In 1986, the logging companies' bulldozers reached our common land," Unga Paren said. The headman gave the impression of being calm and concentrated, but it was evidently difficult for him to speak about the suffering of his people. "We wanted to negotiate with the government and to ask the logging company not to chop our trees down," Unga continued, "but no one would listen to us, and they told us that we had no rights over our land."

When the negotiations failed, the Long Bangan villagers erected a large blockade across the road. It only lasted two days. The logging company called in the police and an army unit, whose response was brutal. A hundred Penan were arrested and put in prison. Their headman spent more than two weeks behind bars.

When Unga was released, the company's bulldozers and chainsaws were already removing the forest trees from Long Bangan collective forest under police protection. The headman was powerless to act, and could do nothing except look on as the finest giant trees from the virgin forest were chopped down and carted off. Poison-dart trees and sago palms—everything was destroyed, even though the loggers were initially only interested in a few easily saleable species, such as *meranti* and *belian* (which is also known as Borneo ironwood).

In the years since, the rainforest all around the village was cleared in stages. All the big trees went first, followed by the smaller ones. The river and ground water were contaminated. There is now no potable drinking water in a place that used to be in the middle of a rainforest: The villagers of Long Bangan are forced to collect rainwater from the roofs in big blue plastic containers, which the logging company gave them as "voluntary compensation".

With the exhaustion of the timber reserves in the surrounding forest, the timber barons shifted their strategy. They no longer waited for the forest to rejuvenate, but preferred to clear-cut the rest of the vegetation as well and to replace the forest with oil palms, which promised to bring them quicker profits. Like a fast-spreading cancerous growth, the oil palm plantations have recently been creeping nearer and nearer to the rice fields around Long Bangan.

What happened in Long Bangan is symptomatic of the tragedy suffered by the native inhabitants of Sarawak in the second half of the 20th century, depriving them of their customary rights and wilfully destroying their unique rainforest culture. However, it also symbolises the dark side of the overall development of Sarawak, which has brought extreme wealth to a handful of individuals, but for others has brought only poverty, disease, and the loss of their most important possession: their traditional land.

Headman Unga expressed deep concern about the future and his fear that the government and logging companies would turn the Long Bangan collective land into an oil palm plantation. On taking leave, I promised him and Aren Umai that the Bruno Manser Fund would not forget their village and would be willing to assist with mapping the collective land, and in drawing up a land rights complaint.

BORNEO'S LAST NOMADS

The helicopter with the Canadian television team landed softly in the forest clearing. It had scarcely taken half an hour to fly from the coastal town of Miri to Long Gita in the upper reaches of the Limbang valley. Ian Mackenzie, a linguist from Vancouver, climbed out, put his hands over his ears to protect them from the noise of the rotor blades, and walked over to the modest wooden huts on the edge of the clearing. The "giant forest demon", as the Penan call their tall friend, had arrived in his home-from-home. He was followed by the camera team, hurrying along behind him. They had come in search of Borneo's last nomads.[29]

Mackenzie has been doing research into the language of the Eastern Penan since 1993. Back then, some 400 of these people were living as nomadic hunter-gatherers in the pristine forest in the northeast of Sarawak. Mackenzie had become fascinated by their unique way of life, since they were the only remaining people in Southeast Asia who did not farm or keep livestock. Meeting the Penan had changed the Canadian scholar's life. He has since spent several months every year in Sarawak, collecting expressions and stories from the culture. Mackenzie has compiled more than 15,000 Penan words in his *Dictionary of Eastern Penan*, a work unique among indigenous language studies in both its scale and sophistication. He has recorded more than 1,300 expressions just for trees and forest plants, showing what a central role nature plays in this rainforest culture.

According to Mackenzie, "farmer" had always been an insult for this people whose traditional livelihood depends on the harvest of wild sago. He suspects that the Penan were not originally nomadic; rather, they may have been farmers who were displaced by war from their settlements and found refuge in the depths of the forest. He goes on to explain that "the Penans discovered that living nomadically in an intact forest requires less work than a sedentary lifestyle."

But as their forest retreat was destroyed by logging, and wild game and sago palms grew ever scarcer, the advantages of their old way of life disappeared. In the first years of the 21st century, the last of the nomads settled down and began to cultivate rice and cassava. "It is a collective decision. They have discussed it among themselves, and determined that their future lies in settlement. Modern life offers opportunities that they want to exploit."

Just off the helicopter, Mackenzie sought out Headman Along Sega, the tenacious nomad and fighter for the Penan cause. For him, this people was the last nomadic group in the Sarawak rainforest. But his search was to end in disappointment. Along too had become sedentary, if only recently; a miserable collection of wooden huts was now his permanent home.

"This is what I feared and expected," said Mackenzie, visibly moved and speaking into the Canadian television camera. "They've settled down here, in secondary forest, in a bunch of wooden shacks. The last time I saw them, they were the world's proudest people. They would not farm. But now they have surrendered and planted rice. They are settled, and it's the end of the ancient nomadic lifestyle."

Less than 120 years after its "discovery" by Charles Hose, the age-old nomadic way of life is close to its end. The Penan "noble savages" in their rainforest paradise have been overtaken by a civilisation that permits no life without money, wage labour or private property.

"I just have to accept the fact that this culture is vanishing," said Mackenzie. "All I can do is build them a tombstone and write an epitaph. But I want that tombstone to be as large as it can be, and I want that epitaph to be a million words long. And I believe I will be writing it for the rest of my life."

Settling down would bring no luck to Headman Along Sega. A few years later, on 2 February 2011, he died of an infected wound in the Limbang town hospital—still poor, and still disappointed in the Malaysian government and in Taib, who had promised the Penan development and prosperity.

"When I am dead, my children will continue the struggle for our land rights and the forest," Along told me when I met him for the last time. Indeed, only four weeks after Along's death, the Penan from the Upper Limbang, along with their Kelabit and Lun Bawang neighbours, filed a land rights claim. The legendary headman did not have the pleasure of experiencing that historic moment. However, his son Menit is one of the plaintiffs. They are demanding that the Taib government recognise their land rights, cancel the logging concessions, and return what is left of the forest to its original inhabitants. Rodney Needham's photographs and his affidavit will be key exhibits in the court proceedings.

Dum spiro spero ("as long as I breathe, I hope") used to be the family motto of the Brookes, the "White Rajahs" of Sarawak. They could not have left a better motto for the country's original inhabitants. For what dies last is the hope for justice and a better future.

THE
WHITE
RAJAHS

In 1841, British adventurer James Brooke
created his own private colony on the island of
Borneo. For a century, his family ruled as kings.
Thirst for oil and geopolitics brought Sarawak into
the British Empire after the end of the Second
World War, and, in 1963, into the new nation
of Malaysia.

A KINGDOM OF ONE'S OWN

The modern state of Sarawak began with James Brooke, the son of a British colonial officer, born in 1803 in what was then British India. At the age of twelve, he was sent to England to be educated, and when he reached sixteen he returned to the Far East as a soldier in the service of the East India Company. Brooke was wounded in the First Anglo-Burmese War and returned to England in 1825.

James Brooke dreamt of building up a network of British trading stations in Borneo. His plan was inspired by the works of Sir Thomas Stamford Raffles (1781–1826), the founder of Singapore and a passionate natural scientist, after whom rafflesia, the giant-flowered parasitic plant, is named.[1] Brooke's opportunity came when he inherited 30,000 pounds upon his father's death, enabling him to buy the 142-tonne schooner *Royalist* in 1836. After various test outings, the *Royalist*, armed with six cannons, set sail from England on 16 December 1838. On 1 August 1839, Brooke landed on the north coast of Borneo in stormy weather. A fortnight later, he anchored in the Sarawak River off Kuching.

At that time, the territory around the Sarawak River belonged to the Sultan of Brunei. The earliest documented use of the name "Serawak" is from the 14th century. It is derived from the old Malay word for antimony, which is one of the mineral resources to be found in the river basin.[2] European sailors feared the northern coast of Borneo on account of Iban pirates, and James Brooke with his well-armed *Royalist* was a godsend in battle against the pirates and helped quash an uprising against the Sultan of Brunei. Thanks to his success in battle, a great deal of political skill and the support of local allies, Brooke gained so much respect in the eyes of the Sultan that he agreed, on 24 September 1841, to transfer governmental power in Sarawak to Brooke in exchange for a modest annual tribute.[3]

From then onwards, Brooke took on the name of "Rajah" and considered himself the head of a sovereign state, which he ruled over with a handful of British officials, and to which he was soon to add two further provinces.

It is said that James Brooke ruled with circumspection, respecting local customs and traditions as far as possible and only intervening when he saw no other solution. That, without doubt, was one of the keys to his success as the founder and ruler of a state. He was particularly concerned about developing trade, combating piracy, and abolishing the head hunting that was still rife at the time amongst the indigenous peoples of Borneo, especially the Iban. During a visit to England in the 1840s, Brooke was knighted by Queen Victoria. He was also appointed Consul-General for Borneo, and his heroic acts were celebrated in the British press.

Sir James Brooke's personalised rule did not, however, meet with the approval of all of the inhabitants of Sarawak. He encountered particular resistance from the indigenous Iban, whose leader, Rentap, became a legend by attacking a fort on the Skrang River belonging to the Brookes. In his counterattack, Brooke shrewdly allied himself with other Iban groups. In 1849, Brooke called in the British navy from Singapore to fight the Iban of Saribas. In the ensuing battle, more than 500 indigenous people were killed.[4] As a result, Brooke was investigated by the British government and accused of disproportionate use of force and atrocities in defeating the Iban.

Though James Brooke might have been militarily and politically successful in his daring exploits, he was not a particularly good administrator or economist. Brooke was able to finance his colonial adventures only by depleting his vast private wealth—even going into debt. At his death in 1868, all the "White Rajah" could bequeath to his successor was—according to Runciman—the "poorly structured and impoverished state of Sarawak."[5]

CONSOLIDATION OF FAMILY RULE

Brooke died without an heir, and without the exceptional talents of his nephew Charles Brooke (1829–1917) his romantic dream of a lucrative private colony would have come to an unspectacular end, with the Sultan of Brunei, the British Crown, or some other foreign colonial power taking control of Sarawak.

Nephew Charles Brooke, however, was a talented administrator: sober, disciplined, and strict with himself and his subjects. He was, in many respects, the opposite of his uncle. Having been brought up in England, he arrived in Sarawak at the age of twenty-three. He then spent most of the next years in the service of his uncle in far-flung outposts of the Sarawak hinterland, alone and without the company of any other Europeans.

When his uncle died, the nephew was determined to continue the family's colonial experiment. The 39-year-old Charles Brooke had himself crowned the Second Rajah, and took over the government of Sarawak.

When he assumed office, it was the beginning of a period of consolidation for the young state. Charles Brooke's biggest success as Second Rajah came in 1888 when he signed a treaty making Sarawak a protectorate of the British Crown. With the British providing defence against foreign powers, the Rajah kept a free hand in domestic matters.

With great diplomatic skill, Charles Brooke continued the expansion of Sarawak that his uncle had begun, and acquired significant new territories in the north from the Sultan of Brunei: the huge Baram region (1881); and the basins of the Trusan, Limbang, and Lawas rivers, so that Sarawak finally surrounded the territory of Brunei. By 1905, the expansion of Sarawak was complete. With a land area of 124,500 km^2 (nearly the size of England) the territory controlled by Charles Brooke on Borneo had grown from the Sarawak basin to a state of respectable dimensions.

For the development of Sarawak's economy, the Second Rajah put his faith almost entirely in local traders and businessmen, and wanted to avoid having European capitalists buying up large tracts of land for plantations. The only foreign trading company permitted to do trade with Sarawak was the British Borneo Company. As vast rubber plantations were being created across Southeast Asia at the end of the 19th century, the cautious Rajah warned against a speculation bubble: "I hate the name of rubber and look on it as a very gigantic gamble [...]." If there had to be rubber plantations at all, then he wanted them to be in the hands of local small farmers.

Charles Brooke opposed the kind of exploitative imperialism that was practised in many other colonies towards the end of the 19th century. Because of this, Brooke gained a reputation as a progressive colonial ruler. Conscious of the atrocities being committed by Belgium under Leopold II in Africa, he declared that "Congo rules cannot be tolerated in Sarawak."[6] He condemned greedy speculation and exploitation by the colonial powers wherever he saw it: "The main consideration should be an honest and upright protection afforded to all races alike and particularly to the weaker ones."[7] By way of contrast, the Rajah determined that the guideline for governing Sarawak should be the "interest of the indigenous people"—albeit as defined by him.

Brooke's position served to legitimise his family's rule—they claimed to govern as mere trustees of the people of Sarawak.[8] At the turn of the 20th century, Charles Brooke and his family were administering their state with around thirty European colonial officers and a number of locally-recruited native officers, the majority of whom were Malays. Charles Brooke likely also feared that the inevitable changes brought by opening Sarawak to the world market too quickly would endanger his own hold on power in the state, which remained fragile at all times.

END OF THE BROOKES IN SARAWAK

This policy of economic and political isolation was continued by Charles Vyner Brooke, the Third Rajah of Sarawak, who took over on his father's death in 1917. To this day, the Brookes still enjoy great esteem in the eyes of many Sarawak inhabitants on account of their prudent governance, and for striking a balance between local interests and their own. Their biggest failing was in education, however, which was totally neglected during their reign. An education department was created at the beginning of the 1930s but was abolished again a few years later for cost reasons. Schooling was left entirely to private groups, such as the various

Chapter 3—Series of photographs

British social anthropologist Rodney Needham took this series of photographs of Penan nomads from the Baram and Tutoh river systems in 1951, during research for his doctoral thesis. They are the earliest known pictures of the Eastern Penan.

missionary schools. Because of this, only very few Sarawak inhabitants ever reached an adequate level of education, and they were poorly equipped for independence when it came.

The 1941 constitution introduced a two-chamber council that would end the Brookes' autocratic regime, and limit the family's monopoly on power. The Third Rajah had already long since lost interest in the bothersome business of ruling Sarawak and was spending more and more time in England. In fact, renouncing his autocratic privileges came easily to him, since he had no son to inherit his throne. Furthermore, he wished to prevent his nephew, Anthony, with whom he was at loggerheads, from being crowned as the next Rajah. Vyner toyed more and more frequently with the idea of ceding Sarawak to the British Crown in exchange for a tidy sum of money—a plan that was also supported by his flamboyant wife, Sylvia.[9]

For the first time, the 1941 constitution had brought the local Sarawak elite—especially the Malays—into the Brooke government. This new form of rule, however, was not destined to last for long. In December 1941, only three months after acceptance of the new constitution, Japanese troops invaded Sarawak, bringing the Rajahs' reign to an abrupt end. The Second World War had spread to the Pacific.

After Borneo was re-conquered by allied troops at the end of the war, the new British government under Clement Attlee came to the conclusion in 1945 that the days of the Brookes were numbered and that Sarawak should come under the direct control of the United Kingdom. The oil fields of Miri and Brunei were regarded as being of strategic importance. Furthermore, the Indonesian independence movement was seen as a threat to British interests in Southeast Asia.[10]

Charles Vyner Brooke was in agreement with the transfer of Sarawak to the British Crown—if he could only secure a lucrative settlement. In addition, there remained the acquiescence of the Malay elite, the Datus, who had gained considerable influence through the 1941 constitution, and who had always been important confidants of the Brookes in ruling Sarawak. The plan was simple: The Rajah's former private secretary travelled to Sarawak in December 1945 in order to secure the agreement of the Datus to cede Sarawak to the British Crown. In his luggage, he had 55,000 pounds sterling to help him achieve his goal. The bribe seemed to work—the last holdout among the Datus was persuaded with a hereditary title, and exclusive rights for the exploitation of turtle eggs on three islands off Kuching.[11]

There seemed no further obstacle to the Rajah ceding Sarawak to the British Crown in exchange for one million Sarawak dollars. However, the announcement that the Brookes were leaving Sarawak triggered a storm of protest. In particular, many Malays were afraid of what would happen to their privileged status in a British colony. Leaked information about the bribes cast doubt on the legality of the whole manoeuvre, obliging the Rajah, who had already settled in England, to return to Kuching for a vote in the new council, according to the constitution. By the end of May 1946, it was finally clear that Sarawak would indeed become a British colony. The Brookes' reign was over.[12]

POLITICS, PROTESTS AND A MURDER

The streets of Kuching remained empty on 1 July 1946 when the yellow, red, and black flag of the Brookes was taken down and the Union Jack raised in its place. The people of Sarawak gave the new rulers an icy reception. No Iban representatives were present, and Datu Patinggi, the head of the Malays, sent a note of apology instead of attending the ceremony with the British Special Commissioner for South East Asia.[13] Moreover, hardly anyone apart from Europeans and Chinese turned up for the evening garden party held in the Brookes' Astana residence with the evening's guest of honour, the legendary World War General and Supreme Allied Commander in Southeast Asia, Lord Louis Mountbatten.

The affections of many Malays and Dayaks still lay with the Brookes.[14] It was their hope that the nephew of the last Rajah, Anthony Brooke, might perhaps still be able to succeed to the throne and reverse the cession of Sarawak to the British. Many worshipped Anthony Brooke in an almost mythical way.[15] So great was their support for him that one of the first acts

of the new British governor of Sarawak, Charles Arden-Clarke, in November 1946 was to impose a ban on Anthony Brooke entering Sarawak.

This was only the beginning of the problems for the British. On 2 April 1947, more than 300 Malay officials working for the Sarawak administration resigned from office as a protest against the new colonial authority, many of them facing considerable financial sacrifices as a result.[16] The popular movement supported by a new Malay intelligentsia was not only directed against rule from London but aimed ultimately at achieving autonomous rule for Sarawak.

At the end of 1949, a young opponent of the takeover stabbed to death the new British governor of Sarawak, Duncan Stewart, just after his arrival. The British responded harshly and the culprit and his supporters were hanged in public. The assassination and the British reaction sent shock waves through Sarawak, and, in combination with intensified police measures by the British, brought about the abrupt collapse of the anti-cession movement.

Early in 1951, Anthony Brooke challenged the cession of Sarawak to the United Kingdom before the Privy Council in London.[17] His challenge was denied, and he responded by formally renouncing his claim to the Sarawak throne, and called on his followers to recognise British rule. Anthony Brooke spent the rest of his life as a peace activist and died in 2011 at the age of ninety-eight in New Zealand.[18]

THE HEADMAN'S LONGEST SPEECH EVER

In January 1962, Headman Oyong Lawai Jau (1894–1968) spoke for more than six hours at a meeting in Long San, a small settlement in the upper reaches of the Baram, the mighty river in northern Sarawak. His guests—headmen and rice farmers from the Baram region—sat on the wooden floor of the longhouse veranda and listened patiently. Sixteen years after Sarawak had been taken over by the British Crown, Oyong had called his people—Kenyah, Kayan and Penan—together to discuss the country's future.

Oyong's alert gaze, the leopard's teeth of his ear ornaments, the brass rings on his extended ear lobes and his traditional Kenyah straw hat gave him—the paramount headman *(temenggong)* of the indigenous peoples of the Baram district—particular charisma. The British colonial authorities also had much respect for the influential indigenous orator, who had proven to be a loyal ally. Towards the end of the Japanese occupation, he had assembled a force of nearly a thousand local warriors and helped the Allies re-conquer Baram district in 1945. In recompense, Governor Charles Arden-Clarke arranged for Oyong to become a Member of the Order of the British Empire (MBE) in November 1947.[19]

The occasion for Oyong's speech was the British government's plan to amalgamate Sarawak, North Borneo (today's Sabah), Singapore and Malaya (today's Peninsular Malaysia) to form a new state to be called Malaysia and to grant them independence. This had been proposed by Tunku Abdul Rahman, the political leader of the Malays and first prime minister of the Federation of Malaya, which had already become independent in 1957. Tunku's "Malaysia Plan" was debated intensively in Sarawak—and met with much disagreement.

Headman Oyong did not hide his displeasure at the plan, but said in the flowery language of the indigenous Baram that he felt that annexing Sarawak to Malaysia was a bad idea. He had words of praise for the British colonial government, which, in contrast to the Brookes, had helped the indigenous people improve their lives: "During the Brooke regime, for the most part we remained ignorant with no one to instruct us or guide us. On the other hand, we were well protected and felt secure. It was only after the war when Sarawak became a colony that the new government showed us how we could improve our way of life, how our gardens and farms could be improved."[20]

He compared the Malaysian peninsula, which had already become independent from the British in 1957, with an orchard, whose trees had grown high and were now full of blossom and fruit, and which was secure behind a sturdy fence made of belian (ironwood). Compared with that, he saw Sarawak as a garden that had been freshly planted and was still young and immature with a flimsy bamboo fence around it. It was Tunku

Abdul Rahman's intention, he said, to move Sarawak's young garden inside his big Malaysian garden and to move Sarawak's flimsy fence inside his strong belian fence. "But although on the face of it this seems a good idea I just cannot agree with him. Why don't I agree with him? I have no need to tell you what happens to a garden when you try to plant trees and shrubs under big trees. They are simply eclipsed by the shadow of the big trees and grow wild. Of course they grow, but they never, to the best of my knowledge, bring fruit. Sooner or later, the only thing to do is to get rid of one or the other. Both cannot survive."[21]

The analogy was unequivocal. Oyong was afraid that the indigenous people of Sarawak would sooner or later be worse off as a result of annexation to Malaysia and would be bound to suffer under the dominance of Kuala Lumpur. He saw the British as the better guarantors of the development of Sarawak than the government of Prime Minister Tunku Abdul Rahman: "I don't mean to despise Tunkus's offer. I only say it won't work as far as we up-river people in Sarawak are concerned. We still need the British to guard our garden, water it, fertilise it, until such time as it is fit to bear fruit. They have much to do still."[22]

What Oyong wanted was neither Sarawak independence nor annexation to Malaysia but for the British to remain in Sarawak for another fifteen years. He stressed that, for him, nothing less was at stake than the survival of his people: "I say all this because I am most concerned about the future of our country and those who will follow us. [...] Our concern must be not for our own selves but for our children and grandchildren. I can have no part with those who could sell us for the sake of a few dollars or a little material gain. I am speaking because of my deep concern for the survival of our race."[23]

The *temenggong*'s views were shared by many of Sarawak's indigenous people. They feared that there was more to be lost than gained through annexation to Malaysia and wanted the British to stay, or pleaded for there be an effective safeguard of their rights. "Unhappy as a child pushed away by his father" is how one Kayan headman from the Baram district put it.[24] Many, however, were at a loss, and did not know what to think of the Malaysia Plan.

A BRITISH BANKER ON A MISSION IN BORNEO

On 19 February 1962, a month after Oyong's speech, a British government commission of inquiry under Lord Cobbold, the former governor of the Bank of England, arrived in Sarawak.

Baron Cameron Cobbold was a thoroughly traditional British gentleman. Born in London and educated at Eton, he had studied at Cambridge and then joined the Bank of England, where he ended his career as Governor, a position he held for 12 years. His world was one of the pound's fluctuating exchange rate, monetary policy, and financial stability for the British Empire. At the end of 1961, he had resigned from his office, and given up his workplace in the heart of the City of London.

The time had come for Cobbold to swap his pinstripe suit and bowler hat for a pith helmet, and to make the journey to Borneo as the envoy of Prime Minister Macmillan. Apart from Cobbold himself, the commission of inquiry had two other British members and two representatives of the Federation of Malaya. Its remit was to establish how the inhabitants of Sarawak and North Borneo felt about the British plan for integrating them in the new state of Malaysia along with the Federation of Malaya and Singapore. Strictly speaking, the outcome of the inquiry was never going to be of any real importance, since the British government had already secretly agreed with Kuala Lumpur in advance of the Cobbold mission that Sarawak and North Borneo would indeed be added to the new state of Malaysia.

The Cobbold commission, however, took its mandate seriously. The team spent two months travelling through Borneo, visiting thirty-five urban and rural centres, where it held fifty hearings. Cobbold's clerk produced a meticulous record of the statements regarding the Malaysia Plan made by more than 4,000 individuals (from 690 groups) who appeared before the commission. The commission also accepted more than 2,200 written opinions.[25]

Those opinions demonstrated the stark ethnic divisions of Sarawak, with its more than forty ethnic groups. The vast majority of the Muslim Malays and Melanaus—who lived in the coastal regions and made up a quarter of the population—were in favour of the Malaysia Plan. They hoped to gain political influence in a new state shaped by Malays under the leadership of Tunku Abdul Rahman. The majority of the mostly urban Chinese—another quarter of the population—were against it, but not unanimously so. The indigenous peoples from the interior, most of whom are Christians, were split. Some, like Headman Oyong, demanded that the British remain, whilst others considered the British withdrawal simply a matter of time. They were coming to terms with the future balance of power.

The Cobbold Report was ready at the end of June 1962. It established that one third of the inhabitants of Sarawak were in favour of joining Malaysia, another third demanded additional safeguards to protect the citizens of Sarawak, while a final third rejected the Malaysia Plan and wanted either an independent Sarawak or a continuation of British rule.

Minds were soon made up in London. Firstly, the Macmillan government was having difficulty in finding the means to maintain an empire in Southeast Asia. Secondly, the United Kingdom was under international pressure to grant independence to its colonies. London was also afraid that Indonesia might try to annex the British possessions in northern Borneo. Under its first president, Sukarno, newly independent Indonesia was regarded as a dangerous gateway for communism in Southeast Asia.

Those fears were amplified still further when, in December 1962, communist-inspired rebels, with support from Indonesia, rose against the Sultan of Brunei and set about bringing the oil fields under their control. British forces were quickly flown in from Singapore and, in a matter of days, forcefully put down the rebellion, which had also spread to neighbouring Sarawak. Twenty armed rebels and seven members of the security forces were killed in the fighting.[26]

The British government was now eager for a pragmatic solution that would result in both independence to its remaining possessions in Southeast Asia and political stability in the region.[27] It declared itself to be in agreement with the Malaysia Plan, arguing that Sarawak, North Borneo, and Singapore would not be viable as autonomous states. On 16 September 1963, they joined with the Federation of Malaya to become part of the new state of Malaysia, with Kuala Lumpur as its capital. Singapore stayed with Malaysia for two years, becoming a sovereign state in 1965.

In June 1963, three months before independence, another election was held in Sarawak, and a first cabinet formed for the new Malaysian federal state. The departing British governor appointed the ministers. The first chief minister was Stephen Kalong Ningkan, one of only a few well-educated indigenous Ibans.

A black-and-white photograph of the transitional cabinet shows ten men wearing suits and ties arranged in two rows in front of the government headquarters.[28] The new chief minister is seated in the front row with his hands resting dutifully in his lap next to the governor, enthroned in the centre. A young lawyer in fashionable clothes stands in the second row staring straight at the camera. He had just been appointed minister of development and labour. His name would soon be heard all over Sarawak: Abdul Taib Mahmud, or simply Taib.

04

SARAWAK'S MACHIAVELLI

In 1963, Taib became a minister in the first cabinet of the Malaysian state of Sarawak. But the ambitious young lawyer had higher ambitions. He teamed up with his uncle Rahman to neutralise Sarawak's non-Muslim indigenous majority. Finally, he managed to ascend to the highest echelons of power.

CONNECTIONS AND COMMODITIES

The 27-year-old political nobody Taib owed his ministerial post in the first cabinet of the new Malaysian federal state of Sarawak to a stroke of fortune—and his uncle. The young law graduate, returning from studies in Australia, had planned to become a judge. A secure job in government seemed just the thing to Taib, who had known poverty as a child. As a deputy prosecutor in Kuching in the service of the British, he was on track for a career as a judge when Uncle Rahman called him into politics.

Abdul Taib Mahmud had been born in Kampung Merbau near Miri on 21 May 1936 as the oldest of ten children of a carpenter employed by Shell. In 1910, the then Anglo-Saxon Petroleum Company—now part of the Royal Dutch/Shell Group—had struck oil on the outskirts of Miri, a few kilometres from the estuary where the Baram River flows into the South China Sea. This had triggered a boom in the small coastal town. People in search of work poured in, some from considerable distances, including Taib's family, which had its roots in the Muslim Melanau community in the small town of Mukah, 200 kilometres to the west.

The family made much of the fact that one of Taib's great-grand-fathers had belonged to the royal court of Brunei, though any fortune had long since been exhausted.[1] To attend secondary school, Taib was thus dependent on help from his mother's younger brother Uncle Rahman Ya'kub—known simply as "Rahman". Rahman had managed to land a job for himself as a civil servant in the British colonial administration, and in 1949 succeeded in organising a scholarship from Shell for his nephew.

Taib learned from a young age that personal connections and money were all he needed to get ahead, and it was these two decisive components that would fuel his meteoric rise to becoming one of the richest men in Southeast Asia.

At the age of thirteen, Taib moved to Kuching, where he successfully completed St. Joseph's secondary school, which was run by Catholic missionaries. A further scholarship through the Colombo Plan for Co-operative Economic Development in South and Southeast Asia took the bright young man to study law at Adelaide University in South Australia, starting in 1956.[2] He would have preferred to study medicine, he later claimed, but his uncle Rahman guided him towards law. Rahman, after all, knew best, since he too had completed a law degree at Southampton in the United Kingdom.

ADELAIDE, MON AMOUR

Taib was happy to be headed for Australia rather than Britain. He had always had a preference for the straightforward Australians rather than the stiff British. For him, Adelaide must have been a dream come true—a dream of education, prosperity, and a cosmopolitan world. It was there that he became acquainted with 1950s American fashion, which had hit the Australian student city—albeit a little later than in the US. In Adelaide he learned to smoke a pipe, which would give him a man-of-the-world aura later in Sarawak. Also in Adelaide he met his first wife, Laila, whom he married in 1959, at a time when they were both still students, and one year later she gave birth to a baby girl.

Taib clearly has some nostalgia for Adelaide and his alma mater. Why else would Taib, who is not otherwise known for generosity, have made donations over the years to the University of Adelaide to the tune of several million Australian dollars? One of the beneficiaries, ironically enough, has been the Centre for Environmental Law.[3] The university has expressed its thanks to Taib by awarding him an honorary doctorate, and, even as late as 2008, named a courtyard the "Taib Mahmud, Chief Minister of Sarawak Court".[4] Taib is also a director of the alumni organisation, Australian Universities International Alumni Convention, which has its seat in Adelaide and is his only official participation in business in Australia.

Adelaide is also the site of the most prominent building in Australia that—officially—belongs to the Taib family: the 18-storey Hilton Hotel, one of the tallest structures in the South Australia capital. It is a popular meeting place for the city's elites—and is valued at over 10 million US dollars. Amongst the organisations that meet there is the Australia-

Malaysia Institute, a body sponsored by the Australian government, tasked with improving relations between the two countries. And the owners of the company running the hotel? Sitehost Pty Ltd.—owned in equal parts by Taib's four children; his wife, Laila, who died in 2009; and an obscure financial entity based in Guernsey, which possibly has Taib himself behind it.[5]

Laila Chalecki, a young refugee from Lithuania, could scarcely imagine the prize she had won in the ambitious young student from Borneo. They soon married, head-over-heels in love. She had been born in 1941, at the height of the Second World War in Vilnius, which was at first under Soviet and later German occupation. After the war, at the age of eight, she travelled to Australia with her parents on board a refugee ship.[6] Her paternal grandfather had been the muezzin of the Vilnius mosque, and a member of the Polish-Lithuanian minority of Lipka Tatars.[7] Her father had served as a young officer in the Polish army and had been one of the lucky few to return home alive from the war. While others were still fighting on the front, in 1943 Abu-Bekir Chalecki enjoyed the rare privilege of studying medicine in Vienna, which was in Nazi hands at the time. After emigrating to Australia, he set up a medical practice and for many years was chairman of the Australian Muslim Council.[8] Laila's mother died in Adelaide in 1952.[9]

Later photographs show Laila with a broad smile standing beside Taib at official functions. She had lived through hardship and uncertainty in early childhood before coming into immeasurable wealth thanks to her husband—who was a good head and shoulders shorter than she was.

They married in the Adelaide mosque on 13 January 1959 (Laila was not yet 18). The brick building in Little Gilbert Street, in the historic centre of Adelaide, is the oldest mosque in Australia. It was built in the 1880s by Afghan camel drivers, who drove caravans across the Australian outback. After the Second World War, the mosque was the place where students from Southeast Asia met Muslim immigrants from war-ravaged Europe. Soon everyone knew everyone else—Taib met the alert Laila, who began studying medicine soon afterwards. When she gave birth to a baby girl in September 1960, she named her Jamilah after her late mother.

In Adelaide, Taib and Laila also met Hijjas Kasturi, a student of architecture who had been born in Singapore, and who, like Taib, had come to Adelaide thanks to a Colombo Plan scholarship.[10] Later, in Kuala Lumpur, Kasturi was destined to become one of the best-known and most successful architects in Malaysia, not least thanks to numerous public contracts from Sarawak. Twenty years later, he built the spacious Demak Jaya Palace for Laila and Taib in Kuching, at a time when they had both long since left behind the poverty of their early years.

Shortly after completing university in 1961, Taib was given his first job as an assistant to a 75-year-old judge, Sir Herbert Mayo, at the Supreme Court of South Australia. One year later, he became the first lawyer from an Asian country to be admitted to practise by the Supreme Court in Adelaide. Laila came to watch when Taib was allowed to try on the judge's wig in the court library.[11]

RETURN TO SARAWAK

Despite all this, the young law graduate from Borneo did not forget his roots. In January 1962, six years after leaving for Australia, he returned to Sarawak with his wife and child, where the debate on independence and the formation of the federal state of Malaysia had begun.

Uncle Rahman was waiting impatiently for his nephew. He needed Taib for his latest project: a new party that would be the vehicle for his political ambitions. The first step, however, was to find a job for Taib. A position as deputy public prosecutor in Miri seemed a perfect fit for the returnee from Australia. Eight years previously, Uncle Rahman had been turned down for a similar post by British colonial officers, but times had changed. Taib got the job.

Barjasa was the name of the Muslim-nationalist party that Rahman had created in 1961, and on whose behalf he now set Taib to work. As employees of the state, the two men found a colleague to push into the public limelight as they themselves worked behind the scenes.[12] The political leadership in Kuala Lumpur had recognised Rahman as a suitable ally and was encouraging him as much as it could.[13] "Rahman and Taib were the favourites of Tunku Abdul Rahman, who later became the first Malaysian prime minister," is how an acquaintance of the family—who wishes to remain anonymous—remembers.

Sarawak had strong Christian roots and a high proportion of non-Muslim indigenous inhabitants as well as many Chinese in its population. As well-educated, pro-Malaysian representatives of the young Muslim intelligentsia, Rahman and Taib appeared to be suitable partners for Kuala Lumpur right from the beginning. Barjasa aimed itself against the British colonial administration and, more controversially, against Sarawak's Chinese too. This is borne out, for instance, by the fact that the Chinese were missing from a list of "Sarawak races" drawn up by Barjasa.[14]

Rahman's assumption of power was planned for mid-1963. The first free elections took place between April and June 1963, still under British supervision, before Sarawak was granted its independence.[15] Rahman stood for election and hoped he would then become the first chief minister in the transitional cabinet, whose job it would be to steer Sarawak from a British colony to a Malaysian federal state. That, however, is not how things turned out. Instead of winning the election, Rahman suffered a crushing defeat. It was the Iban—the largest indigenous group in Sarawak—who claimed the right to fill the post of chief minister.

It became necessary to resort to a plan B. The big moment in Taib's life had come. The outgoing British governor of Sarawak decided to make use of his right to nominate three members of the state assembly, but was determined not to include any election losers. So instead of Rahman, he placed the latter's nephew, Taib, on his list.[16] The intention was that Taib would represent the Muslim intelligentsia, and, at the same time, he was given the post of minister of communications and works.

On 22 July 1963, two months before the founding of Malaysia, Taib was sworn into office.[17] A historical upheaval, encouragement from his uncle, and a fair measure of luck had lifted the 27-year-old to a position of power. And yet he quickly became accustomed to that power, and soon set about increasing it.

"Proud and conceited" is how Taib and Laila are described by contemporaries who knew the young pair in Kuching in those days. Taib proudly displayed his European wife in Sarawak and was one of the first people in Kuching to own a car—a yellow Opel Rekord. On top of his foreign education, rapid political success soon went to Taib's head. The obsessively ambitious young politician hated being in his uncle's shadow—whose charisma, popularity and oratory skills he was never going to match. After suffering electoral defeat in Sarawak, Uncle Rahman resumed his political career in Kuala Lumpur, where it did not take long for him to become a minister himself.

"Contrary to Taib, Rahman had the skill to stir the emotions of his listeners and to move them to tears," is how one Malay from Kuching remembers him. "He also knew the Quran much better and was thus also prayer leader in the grand mosque in Kuching, which was something that Taib was never going to become."[18] While Rahman was a man about town, who maintained a large circle of friends (and had numerous love affairs), Taib remained an elitist; he concentrated his social life on those contacts that were politically useful for him.

If Taib lacked friends, he made up for it by cultivating his family circle. After the death of his father, Taib, as eldest son, assumed responsibility for his nine younger siblings, of whom his youngest brother, Arip, born in 1957, was only three years older than Taib's own daughter. Out of his salary as a minister, Taib gave carefully calculated pocket money to his youngest brothers and sisters. In return, he expected obedience—and occasionally even resorted to the rattan cane when they failed to abide by his instructions.[19]

With just one brief hiatus at the end of 1967, Taib remained a minister for five decades, either in Sarawak or in the Malaysian federal government, until leaving the ministry to become Governor of Sarawak in March 2014. Over this time, he claimed far-reaching powers, gathered legendary wealth, and, many argue, was the centre of a culture of corruption unequalled in Malaysia. He is the only politician in Malaysia whose career has spanned the period from independence in 1963 to the present day. He has seen six prime ministers come and go, but he alone has remained.

A YOUNGSTER IN THE FIRST SARAWAK CABINET

The long-awaited day came on 16 September 1963. A new nation was born with pomp and ceremony in Kuala Lumpur. Sarawak and Sabah along with Singapore and the twelve states of the Federation of Malaya joined to form the new state of Malaysia. "I pray that God may bless the nation of Malaysia with eternal peace and happiness for our people," said Premier Tunku Abdul Rahman, the founding father *(Bapa Malaysia)* of the new state in his "Malaysia Day" speech. The day before, the last British governor, Sir Alexander Waddell, had taken leave in a short ceremony in the Sarawak capital of Kuching and boarded a ship for London.[20] It was the last time that his white uniform with feather-decked tropical helmet would be seen in Borneo.

The new chief minister of Sarawak was Stephen Kalong Ningkan, of the indigenous Iban. The 43-year-old politician was one of only a few educated Iban with any administrative experience. His previous professional career had been with the police, as a teacher, and in a hospital run by Shell in Brunei; he had also spent some time in London. In 1961, he had been one of the founders of the Sarawak National Party (SNAP), the majority of whose members were Iban, along with some Chinese, and he had become its secretary general.[21] It was the outcome of a political compromise that an Iban, in the person of Ningkan, became chief minister, while a Malay was given the office of governor.

As minister of communications and works, Taib concentrated on the construction of new roads through the rainforest state, which until then had remained generally untouched by roads. Many years later, Taib still goes into raptures when recollecting those glory days. He stresses his achievement in increasing annual road building from 25 to

60 miles within two years.[22] And yet, he did not really enjoy his first ministerial office in the Ningkan government. He was frustrated being the youngest minister, often not drawn into consultation until a decision had long since been made and only needed rubber-stamping by the cabinet. That offended Taib's pride, since he believed that, with his university degree, he deserved more respect. Ningkan very much preferred to seek the advice of other members of the cabinet rather than Taib's. In matters of important business he would turn to the two Chinese, James Wong and Teo Kui Seng; and the three Brits, who, as experienced civil servants of the former colonial administration, occupied the key positions of minister of finances, secretary of state, and attorney general.[23]

There can be no doubt that, despite independence, a crucial element of power in Sarawak remained in British hands throughout the initial years. One factor contributing to that was the presence of British army units in Sarawak, whose stated function was to repel communist rebels along the long border with Indonesia. Indonesia's President Sukarno had been opposed to the creation of Malaysia, and, from 1963 onwards, he attempted to destabilise the new Malaysian states of Sarawak and Sabah, with the ultimate aim of bringing them under his control through so-called *konfrontasi* (confrontation). He failed in that, however, and on 11 August 1966 a peace treaty was signed in Jakarta between Malaysia and Indonesia.

The government in Kuala Lumpur, like Taib, was anything but happy with the situation in Sarawak, where power was held by an alliance of Iban and Chinese under British influence, with the Malays excluded to a large extent from the most important decision-making processes. It caused annoyance, for instance, that Stephen Kalong Ningkan insisted on keeping English as the official language and rejected its replacement by Malay as the only official language.[24] In a public statement in 1964, Ningkan even went so far as to state that British troops should remain stationed in Sarawak on a permanent basis.[25] His government was also accused of being under the influence of Chinese business people.

A COUP

With support from Kuala Lumpur, Taib and his uncle Rahman now waited
for an opportunity to topple Stephen Kalong Ningkan and set up a Malay-
Muslim coalition government along the lines of the one in West Malaysia.

In 1965, when Ningkan announced plans to introduce a new land law
that would have allowed Chinese to acquire indigenous land, Taib seized
the opportunity and tried to overthrow him with the support of prominent Ibans who were afraid of losing control over their land. However,
hot-tempered Ningkan promptly sacked Taib. Under pressure from Kuala
Lumpur, Ningkan was forced to reinstate Taib in the cabinet two weeks
later, but Taib was humiliated by being placed at the lowest rank in the
cabinet.[26]

If Taib had not already decided to seek revenge, he soon made plans to
do so. He was supported by Premier Tunku—who disliked British influence in Sarawak—as well as by his uncle Rahman, who by then had risen
to an influential position in the federal government as a deputy minister.

Taib and Rahman lobbied members of the Sarawak state assembly,
and collected signatures for a vote of no confidence against Chief Minister Stephen Kalong Ningkan. Politicians were summoned to Kuala Lumpur one after another and pressured into signing. Along the way, Taib
skilfully exploited the rivalry between various Iban groups and entered
into an alliance with Pesaka, a group competing with Ningkan.

In mid-June 1966, Prime Minister Tunku announced that Chief Minister Ningkan had been dismissed; the majority of the members of the ruling coalition had lost confidence in him. The Minister for Home Affairs
and Justice was dispatched to Kuching and immediately appointed a new
Chief Minister of Sarawak: Tawi Sli, an Iban and a member of the Pesaka
party—he was considered a compliant subject of Kuala Lumpur. The
British in the cabinet, source of a dispute between Ningkan and Tunku,
were dismissed.

However, Ningkan refused to resign. He argued that only the Sarawak
State Assembly had the authority to decide on a motion of no confidence.
He took court action against the decision from Kuala Lumpur and stub-

bornly defended his position and Sarawak's independence: "If the Prime Minister thinks that, with the help of his puppets, he will succeed in making Sarawak a colony of Malaysia, then Tunku is suffering from a terrible illusion."[27]

In early September 1966, the Kuching High Court found in favour of Stephen Kalong Ningkan and ruled that his dismissal had been unconstitutional. Reinstated, Ningkan declared all government decisions taken since his dismissal null and void. He called a general election.

For Sarawak to have so much independence was going too far for Kuala Lumpur. In the escalating power struggle, Premier Tunku decided against the rule of law and democracy. This would be a test case of his federal powers; Kuala Lumpur would settle the political crisis. Using the pretext of a threat to public order, the Malaysian government declared a state of emergency in Sarawak on 15 September 1966, amended the constitution, and then gave the governor the right to convene a sitting of the Sarawak State Assembly. Under pressure, a thin majority against Ningkan had been scraped together in the meantime, and the chief minister was dismissed once more.

The message was clear: anyone not willing to bow to Kuala Lumpur's will would be dismissed. Still, Chief Minister Ningkan refused to yield. And when he finally did leave, he boasted "I shall return!" and "I shall never be a party to Malay domination."[28] History proved him wrong, however; he never regained power, languishing as leader of the opposition until withdrawing from politics in the 1970s.

LORD OF LOGGING

Premier Tunku had every reason to feel pleased. Though Singapore had left Malaysia in 1965 on account of ethnic tensions between Malays and Chinese, Kuala Lumpur had finally brought Sarawak under its control, and no one was going to change that. The British officers had lost their power.

The big winners from Ningkan's dismissal were Tunku's protégés: Uncle Rahman and Taib himself. One thing was clear: No one would get anywhere in Sarawak without their support. And, together with Premier Tunku, they were already eyeing a larger prize: Kuching itself.

While Uncle Rahman remained in his ministerial post in Kuala Lumpur, Taib, in Sarawak, grabbed the strategically important Ministry of Agriculture and Forestry, into which he soon absorbed the Development Department also. Furthermore, he had risen to the rank of Deputy Chief Minister, number two in the Sarawak government. Now was the time when Taib truly blossomed. Finally, he was lord over the Sarawak logging concessions—a powerful position indeed. Virgin forest covered an area more or less the size of England, full of giant trunks worth billions of dollars. Hundreds of thousands of hectares of ramin, belian, meranti, and the other precious timbers lay waiting to be harvested.

Not everyone yet recognised the true value of Sarawak's forests; Taib did. He also grasped, however, that he had to take the long-term view and proceed with great caution. This was because the Iban still held much power.[29] Chief Minister Tawi Sli was considered weak, but, as a representative of the largest population group in Sarawak, he was a danger to Taib's ambitions.

Photographs taken at that time show Taib as a flamboyant young politician, with an American hat and dark sunglasses, travelling everywhere in Sarawak, visiting remote communities and giving speeches. The pipe-smoking politician had a much more modern appearance than any of his contemporaries. He would walk through swamps in his bare feet, while others wore shoes. He let children hang garlands round his neck, and smiled confidently into the cameras. Now that he had gained in stature, he could afford to show a playful side. But despite this nonchalance, he took care not to antagonise his conservative rural voters. At religious events, he dressed carefully in traditional Malay clothes. He waited before undertaking a pilgrimage to Mecca, so as not to appear pretentious or ostentatious.[30]

Taib was successful with his political core, at least amongst the Muslim Melanaus and Malays, who formed his political base. He and Rahman succeeded in uniting Sarawak's Muslims—who had long been at loggerheads—by bringing them together into a new party called Bumiputera

("Sons of the Earth"). Taib became party secretary general at the end of 1966. "Unity is strength," was his clear message to his supporters, who were aware of their status as a minority relative to the Iban. Taib was viewed with greater scepticism by the non-Muslims, in particular by the Chinese, many of whom regarded him as anti-Chinese and considered him Kuala Lumpur's mouthpiece.[31]

We do not know precisely what transactions Taib conducted with Sarawak's timber resources from summer 1966 onwards. Nor do we know the sum of his own profits from them. It is likely, however, that by the end of 1967, Taib was already the richest politician in Sarawak. And it was also becoming clear that Taib's new form of political economy was going to be lethal for Sarawak's forests.

THE TIMBER CURSE

"Timber is, in a way, our greatest political curse," said Donald Stephens, the first chief minister of Sabah in 1967.[32] In the early 1960s, a model of party financing was developing in Sabah that would come to be the norm in Southeast Asia. It has cost the state coffers immeasurable tax short-falls, quite apart from the ecological damage it has caused. The American economist, David Walter Brown, has estimated that between 1970 and 1999, over 25 billion US dollars in timber revenues were "unofficially appropriated" by timber concessionaires, heads of states, and their prox-ies and clients in Sarawak and Sabah. Instead of feeding the public treas-uries, this money went into the pockets of a handful of politicians and timber tycoons.[33]

The principle is as simple as it is criminal. Politicians in government hand out logging concessions to their favourites and, in return, pocket bribes, which they use to finance their electoral campaigns—as well as for private purposes. Given that there are virtually no other sources of fund-ing for the political parties—and that electoral campaigns are extremely costly in the remote rural regions of Borneo—those in office have a de-cisive advantage over challengers. In Sarawak, whoever managed to gain control of the logging concessions was in a position to harvest enormous sums of money and win every electoral campaign by a wide margin, making it virtually impossible to force them from office. "Stability through corruption" was the magic formula in politics on Borneo. The price was the loss of the rainforests—and the death of democracy. This same method of financing was practiced in the neighbouring Philippines by Ferdinand Marcos, who ruled there from 1965 until 1986, firstly as its elected president, and then as dictator.[34]

In contrast to Sabah, where logging concessions had already played an important political role prior to independence, Sarawak was lagging behind and "the top politicians were still not quite so aware of the potential of logging licences as were their counterparts in Sabah."[35] Even in Sarawak, however, astute business people had already found ways of securing logging concessions during the British colonial period, and had thus obtained a significant influence over government. The most prominent example is James Wong, deputy chief minister in Stephen Kalong Ningkan's first cabinet and one of the most influential Chinese politicians in Sarawak right through the 1990s. His Limbang Trading Company had obtained a sizable logging concession for the virgin forests in the upper reaches of the Limbang River in 1949. It would make him a multimillionaire.[36]

What doubtless upset Taib was that most of the licences were in the hands of Chinese business people on whom he could not rely to provide him with political support. So he took a daring step in order to stop the flow of money to his political opponents; he simply froze all logging concessions in Sarawak.[37] His pretext was a United Nations' Food and Agriculture Organisation (FAO) study examining the long-term use of Sarawak's timber resources. "One of my most difficult tasks was to plan the proper exploitation of our forest resources," is how Taib later justified his actions. "To do this, I had to freeze all forest licences and later reissue them. No one liked it but I had to do it in order to convince the FAO."[38] Whether one believes this statement is immaterial—from that day forth, if anyone was going to pocket bribes from the reissuing of logging concessions, it was Taib and his uncle.

For a year, everything went well in the new cabinet that was officially led by Tawi Sli, but controlled behind the scenes by Taib. Then the situation worsened; Taib's Iban partners demanded Taib's resignation. They accused him and his uncle in Kuala Lumpur of only working for Malays, and of neglecting the non-Muslim indigenous peoples.[39] On top of that came accusations that he had misused timber levies to finance mosques and prayer rooms.[40] The poor finances of the Iban party, Pesaka, also played a key role in the affair; Taib had blocked the issuing of two big logging concessions to Pesaka favourites.[41]

Taib had taken things too far and, for the second time, was ejected from the cabinet by an Iban chief minister. This time he really did have to go, but in Kuala Lumpur, Premier Tunku Abdul Rahman already had a new post lined up for his reliable protégé from Sarawak. It did not take long to clear a seat for Taib in the federal parliament, and he was able to start a ministerial career in Kuala Lumpur.

THE GRAND COUP

A holiday spent in Hong Kong in December 1967 gave 31-year-old Taib the opportunity to reflect on his professional ambitions. Should he really stay in politics, with the relentless fight for power and influence, where he had few allies and many jealous enemies? Or would it not be better for him to embark on a career in business and set out to make a fortune? "He felt he had been abandoned by the people he thought were his friends," is what Taib's then spokesman and fawning biographer, James Ritchie, wrote about that time.[42]

Fortunately for Taib, Premier Tunku and Uncle Rahman were still in Kuala Lumpur. They—especially Tunku, who still had things very much under control in the Malaysian capital—promised to help. Taib and Laila moved to Kuala Lumpur with their 7-year-old daughter, Jamilah, and 4-year-old son, Mahmud Abu Bekir, and Taib took up the post of assistant minister of commerce and industry in March 1968.

Tunku was impressed with Taib's talent, and also with the fact that the young politician was still very pliable and receptive to the premier's ideas. What counted above all, however, was Taib's unswerving loyalty. "I consider him one of my best friends," the legendary *Bapa Malaysia* wrote of Taib in 1990, shortly before the premier's death.[43] Even at a ripe old age, whenever Tunku paid a visit to Kuching, Taib would travel with him in the aircraft and would push his wheelchair in person.

In contrast, we know very little about the relationship between Taib and his uncle Rahman during the time they were together in Kuala Lumpur. It is probable that the feelings between the two were a mixture of friendship and mounting rivalry. Observers say that the self-centred, calculating Taib, who regarded himself as more intelligent than his uncle, developed an inferiority complex vis-à-vis the charismatic Rahman. Two decades later, it would come to an open conflict and power struggle between the two. But in those days a shared project bound them: the political neutralisation of the Iban and the assumption of power in Sarawak.

Their opportunity came with the pending elections, which had originally been scheduled for mid-1969 but were postponed by a year on account of the racial conflicts in West Malaysia. The violent clashes between Chinese and Malays in Kuala Lumpur on 13 May 1969 came as a national tragedy for the young state, and meant the abrupt end to a short period of political freedom. Sarawak was not affected by the clashes, and the legendary Sarawak Rangers, an army contingent of Iban soldiers, were sent into Kuala Lumpur to calm the situation.

The clashes gave the government in Kuala Lumpur the welcome excuse to take power from Sarawak's Iban chief minister, Tawi Sli. All across Malaysia, the most important government powers were transferred to a "State Operations Committee" (SOC), but in Sarawak, exceptionally, it was not the chief minister who was placed at the head of this body, but a former federal civil servant, a Malay. One of the first official actions of the SOC was to freeze the logging concessions once again and thus end the flow of money to the Iban politicians and their Chinese financial backers just ahead of the general election.[44]

The big coup succeeded a year later, after the 1970 elections. Although they won only a quarter of the seats in the newly-elected state assembly, Rahman's and Taib's Bumiputera party managed, through canny manoeuvring, to land the post of chief minister for Rahman. The decisive factor was the sudden change of sides by the Chinese opposition party, SUPP, which was hoping to gain more from supporting Rahman than from yet another Iban head of government.

What happened in Sarawak was a carbon copy of the political model of Peninsular Malaysia: a government under Muslim leadership supported by a subservient Chinese partner. The Iban were fobbed off with two ministerial posts; the period of their political dominance had ended. Now power would sit in the hands of the Melanau ethnic group—in the persons of Rahman and Taib—, itself a tiny minority amongst the Sarawak Muslims, constituting less than 5% of the state's population.

No one would have imagined that, when Rahman moved into office in July 1970, it represented the founding of a family dynasty that would hold on to power in Sarawak for more than forty years—right up to the present day. Without strong support from the federal government in Kuala Lumpur, these two Melanau politicians would never have been able to manoeuvre themselves into such a position. Moreover, without the widespread logging of Sarawak's rainforests to finance their policy and their profligate lifestyle, they would never have been able to gain such dominance over politics and business in Malaysia's largest federal state.

PETRODOLLARS FOR KUALA LUMPUR

The tiny Sultanate of Brunei, one of the richest countries on earth, lies on the north coast of Borneo, a few kilometres to the north of Miri, the town where Taib was born. It has a surface area of only 5,765 square kilometres. On its inland side, Brunei is totally surrounded by the much larger Sarawak, and Brunei's coast is a natural prolongation of Sarawak. Brunei's 400,000 inhabitants pay no taxes, medical care is free for them and, if suffering serious health problems, they have the right to fly to

Singapore 1,250 kilometres away to be treated—paid by the state. The public debt is 0% of gross domestic product. While Western tourists regard Brunei as rather tedious—among other reasons because it is forbidden to serve alcohol in public under the country's strict Islamic law—Brunei's tropical rainforest is stunning. In contrast to Sarawak's, it is comprised almost entirely of old-growth trees.

Anyone travelling on the Pan Borneo Highway from Miri to Brunei is bound to notice the difference almost instantaneously. The road crosses the mighty Baram River on the Sarawak side of the border, with vast tracts of land in the river's catchment laid bare by the logging industry. Sludge oozes into the estuary in Kuala Baram, where large terminals await the barges laden with timber. Tropical timber worth billions of dollars has already been shipped from here to foreign countries—above all to Japan, but in recent years in increasing amounts to South Korea, Taiwan and China too.

Just beyond the border comes the Belait River, which flows through Brunei territory over the whole of its length. The crystal-clear water reflects the old-growth trees, and, behind the river banks, there are vast peat swamp forests, home to rich biodiversity. This is what Sarawak's forest must have been like only a few decades ago.

Many Sarawak citizens wonder if their state could have become a second Brunei, if, like Brunei, they had remained independent from Malaysia in 1963. Oil and gas revenues would have remained in the country instead of being siphoned off to Peninsular Malaysia. A mere 5% of the royalties paid for the crude oil and natural gas extracted in Sarawak (and in the neighbouring state of Sabah) remain in Borneo, while the rest flows into the federal coffers in Kuala Lumpur, which, true to established colonial practice, has been growing rich from exploitation of the natural resources of its far-flung territories.

It is true that the oil field discovered near Miri in Sarawak in 1910 turned out to be much less abundant than the oil fields in nearby Brunei, and had already been more or less exhausted by the end of the Second World War.[45] However, the discovery of new offshore crude oil and natural gas reserves on the continental shelf just off Sarawak from 1968 onwards triggered a conflict between Kuching and Kuala Lumpur. When negotiations failed,

Kuala Lumpur seized the disputed parts of the continental shelf invoking the 1969 emergency measures.[46] When the question came up again in the early 1970s, Taib, as minister of primary industries, was the responsible member of the federal government, and his uncle Rahman was chief minister of Sarawak. The conflict was sorted out as an internal family affair.

Even Rahman, however, was not willing to accept his nephew's more extreme proposals, including the adoption of a hydrocarbon bill ceding the rights to Sarawak's oil and gas reserves to the federal government in Kuala Lumpur—without compensation. Rahman was appalled, and in such a rage over his nephew's suggestions that he urged the Sarawak attorney general to protest to the federal government and took legal steps to stop the planned changes.[47]

Taib was forced to give up his plans, and Kuala Lumpur sent a new negotiating team to Sarawak. In the end, Rahman secured a 5% royalty on crude oil and natural gas extracted in Sarawak. Yet, importantly, Sarawak lost the power over its own oil and gas reserves, smoothing the way for the Petroleum Development Act of 1974, and the creation of Malaysia's state oil company, Petronas.[48]

The new law reversed the balance of payments between the federal government and Sarawak. Up until then, Kuala Lumpur had provided financial support to the underdeveloped federal state on Borneo, but now Kuala Lumpur was able to fill its pockets with the petrodollars from Sarawak. Taking just the six-year period from 1974 to 1980, more than a billion ringgits (roughly US$ 460 million) in royalties and taxes from the production of oil and gas in Sarawak were paid to the federal government.[49] When a huge liquid gas production facility went on-line in Bintulu in 1983, the revenues rose still further.[50]

SULTAN OF SARAWAK

Taib was a minister in Kuala Lumpur for thirteen years before Uncle Rahman recalled him to Sarawak. Thanks to the various portfolios he had held in the federal government—not just in primary industries but in planning, defence, and information as well—, Taib had acquired wide-ranging experience. He seemed predestined for the highest government position in Sarawak. After suffering a heart attack in October 1980, Rahman felt that the time had come for him to relinquish his office to another member of the family.

A by-election gave Rahman the opportunity to get Taib back into the state assembly, and then to appoint him as a minister. Soon after, Rahman resigned as chief minister and had himself appointed governor. The manoeuvre succeeded, and the federal government in Kuala Lumpur, which had hoped to stop a Taib family dynasty in Sarawak by nominating a candidate of its own, found itself confronted with a fait accompli.

On 26 March 1981, a few weeks before his 45th birthday, Taib ascended to the office of chief minister of Sarawak. From the onset, he strove to rule alone and showed no intention of allowing anyone to meddle—not his uncle, not Kuala Lumpur and most certainly not the Iban leaders, whom he downgraded to minor members of his retinue. As his deputy, he chose Alfred Jabu, an acquiescent Iban politician from Betong who proved susceptible to bribery. Jabu served as proof of Iban representation and yet would never be a threat. Taib, once a promising beneficiary of a Colombo Plan scholarship, had become a Machiavelli.

BLOWPIPES AGAINST BULLDOZERS

As chief minister of Sarawak, Taib presided over the state-wide clearing of rainforests. He kept a firm grip on power and permitted no one to obstruct him. Acts of resistance from the indigenous peoples were met with police violence. The Swiss environmentalist Bruno Manser spent fifteen years fighting against Taib's policies as he championed the rights of the Penan. And then he disappeared in the rainforest without a trace.

BROTHERLY LOVE

Initially, Taib was unable to consolidate power effectively. His uncle Rahman held on to the governorship while key positions in the cabinet were staffed by Rahman's allies. The issuing of logging concessions, in particular, was still not fully under Taib's control, as one of his uncle's acolytes retained the key position of forestry minister.

Taib, therefore, decided to play it safe, and, halfway through 1983, he sent his brother Onn to buy real estate in Canada. The first few million dollars amassed by the chief minister were better invested abroad, away from prying eyes. In Ottawa, Onn Mahmud and two of Taib's children, Jamilah and Bekir, jointly established the Sakto Group on 2 September 1983.[1] Since that date, the company has handled transactions totalling several hundred million dollars (see chapter 1).

Having completed his business in Canada, Onn's next port of call was on the other side of the Pacific, namely in Hong Kong, where he set up the Regent Star Company on 22 November 1983. It was destined to become the chief clearing house for timber kickbacks paid into the Taib empire.[2] The purported director and shareholder of the new company was a Chinese employee, whom Onn pushed into taking part. That very same day, Onn also founded the investment company Richfold Investment Ltd. The two companies shared an office on the tenth floor of a characterless concrete block in Connaught Road in the financial district of Hong Kong. While the Taibs were not officially connected with Regent Star, ownership at Richfold Investment was clear. Of the 50,000 shares, 49,999 belonged to Taib's brother Onn and just one, pro forma, to his employee, Shea Kin Kwok.[3]

More than twenty years later, tax authorities in Tokyo found that the shipping companies that had exported tropical timber from Sarawak to Japan had paid sums to Regent Star running into millions. The conditions for timber exporters were clear; without kickbacks to Regent Star, there would be no export permit. It was absolutely essential to keep Onn in a favourable mood, since his brother, Taib, had awarded him control over the privatised tropical timber export agency, Dewan Niaga. At the time of writing, the agency continues to have a key role in the Taib family empire.

And yet the true ingenuity, as far as Dewan Niaga is concerned, lies not in the official business, but in what goes on behind the scenes. Whoever has control over the export of timber is also in a position to manipulate the export statistics. Accordingly, the Taibs can be sure that they are the only ones who grasp the true value of timber exports from Sarawak so they are able to pocket kickbacks for every log moved out of the country.

Onn has been richly rewarded for the business conducted in the service of his brother. With assets totalling an estimated two billion US dollars, he remains the number two in the Taib family. His portfolio includes plantations worth hundreds of millions of dollars, a luxury resort on the tropical Malaysian island of Langkawi, as well as hotels and other properties in Singapore and Australia.[4] Onn hit the headlines in Australia in 2013 after paying no taxes on real estate worth a hundred million Australian dollars. A former business partner accused him of having siphoned all the profits through offshore trusts on the Cayman Islands and the Isle of Man.[5]

TROPICAL TIMBER BONANZA

Taib's assumption of office was marked by a radical acceleration in the rate of clearance of Sarawak's rainforest. Vast areas of forest were cut down in only a few years, and logging advanced from the peat swamps near the coast to the interior of the country, where the virgin forests of several *Dipterocarpaceae* species held the promise of speedy wealth for the loggers. Ramin timber had already become rare by then, so the timber barons threw meranti and belian trunks onto the international market instead. The largest part of this timber made its way to Japan, where the booming construction industry had developed an insatiable appetite for tropical hardwoods. The next most important export markets were Taiwan and Korea. At the end of the 1980s, these three Asian countries together absorbed nearly 90% of the timber exports from Sarawak.[6]

Back in 1967, when Taib was forestry minister, a study had been commissioned by the United Nations Food and Agriculture Organisation (FAO) on forestry in Sarawak. It took until 1972 for it to be published, and its central recommendation was to limit the volume logged annually to 4.4 million cubic metres in order to ensure sustainable use of the forests.[7] Taib and his uncle had, however, lost all interest in limiting the timber harvest once the Iban politicians had been stripped of their power (see chapter 4). Whereas Sarawak's timber production had still been around 2.3 million cubic metres in 1965, it swelled during Taib's uncle Rahman's eleven years in office from 4.7 million cubic metres in 1970 to 8.8 million cubic metres in 1981. Taib, however, planned to extract much more than that. As early as his second year in office as chief minister, more than 11 million cubic metres of timber were felled.[8]

Given the rapid destruction of the rainforest, Taib came under political pressure. So, at the end of 1989, he allowed the International Tropical Timber Organisation (ITTO) to carry out a new analysis of Sarawak's timber industry. Perhaps looking for a realistic compromise, the experts increased the maximum timber harvest for "sustainable" use of the forest to 9.2 million cubic metres.[9] Taib flatly ignored that recommendation as well. In 1991, the logging companies took 19.4 million cubic metres of timber from Sarawak's forests—more than four times the volume recommended by the FAO and twice the timber harvest recommended by the ITTO. It took only a few years for Sarawak to become the world's largest exporter of tropical timber.

That, however, was more than the forests were capable of yielding. Timber production in Sarawak has been falling continuously since 1992, and with 8.2 million cubic metres in 2013 it reached its lowest level in thirty-five years.[10]

During Taib's first six years in office, members of his family and his political allies received logging concessions for 1.6 million hectares. In addition to that, 1.25 million hectares landed in the laps of his uncle's supporters.[11] This meant that logging rights for 2.85 million hectares of tropical forest were in the hands of a single family. The value of these logging concessions was enormous. It has been estimated that those held by Rahman alone were worth 9 billion US dollars in 1987. If the value of Taib's own concessions is added to that, then timber reserves in value in excess of 20 billion US dollars were assigned to a handful of beneficiaries within only a few years.[12] Millionaires were made overnight.

It was a tropical timber bonanza for Sarawak's six large tropical timber conglomerates. The "big six"—or alternatively, the "dirty six"—all went on to become global players: Rimbunan Hijau, Samling, KTS, WTK, Shin Yang and Ta Ann.[13] They all owed their rapid growth more than anything else to good relations with the new chief minister.

This is most striking in the case of Ta Ann. The company was founded in the mid-1980s by Taib's cousin, Hamed Sepawi, and two of his partners; in only a few years, its volume of business skyrocketed. It had become a multinational timber conglomerate through cronyism alone. Another company with a notorious reputation on account of its political connections was Limbang Trading, which controlled logging rights over 124,000 hectares of virgin forest in the remote Limbang Valley in Sarawak's north. Its proprietor, James Wong (1922–2011), a prominent Chinese businessman and politician, held the office of minister of the environment in Taib's government from 1987 to 2001. There could hardly have been a better springboard for conducting lucrative logging business.[14]

COCA COLA DIPLOMACY

Harrison Ngau was 17 years old when he had his first contact with the timber industry. Loggers employed by WTK (the company's name is comprised of the initials of its founder, Wong Tuong Kwang) made their way into his village, Long Keseh, on the Baram River in the north of Sarawak in 1976. At that time, the upper reaches of the Baram were still covered in virgin forests, almost exactly as they had been at the time of Charles Hose's research expeditions at the end of the 19th century (see chapter 2).

"If you are scared, they eat you alive. If you are brave, maybe you have a chance. So you'd better be brave," Harrison Ngau told me in 2012. In using the word "they" he was referring to the timber barons and the Sarawak government. The affable lawyer was born in 1959. He related the story of his life in a calm voice, sitting beneath a map of the world in his office on the outskirts of Miri.[15]

Harrison Ngau is an icon in the Sarawak human rights movement. He is one of very few to have campaigned without interruption since the 1980s for the rights of the indigenous peoples and the protection of Sarawak's tropical forest—initially working for an NGO and later as a lawyer. His people, the Kayan, used to be much-feared headhunters.

"I had just completed my second-to-last year at secondary school in Marudi and had gone back to my village for the Christmas holidays when a large meeting was convened in the longhouse. The reason was that a timber company had been granted a logging concession for the land behind our longhouse. We knew nothing at all about it until WTK suddenly turned up with bulldozers, heavy machines and chainsaws. The people in our longhouse wanted to stop the loggers, but since many of them had never been to school, I helped them write letters to the company."

As the meeting was taking place, the people from WTK turned up with a pile of things to eat—biscuits and Coca Cola. "Everyone started to eat, and nothing more happened. As time went on, it dawned on me that nobody was going to speak. Since I was the one who had drafted the letters, I would have to explain my case, and, though just a young lad, I would have to say that the request to stop chopping down the trees was supported by the entire village." The people of Long Keseh had been subdued. It did not take long before their resistance collapsed completely and they began to seek handouts from the timber company. "I was shocked. In the past, our people had been strong and courageous, yet now they were yielding to the slightest pressure from outside and surrendering their land to the loggers. I thought to myself: 'If our leaders are so weak, then there are huge problems coming our way.'" It transpired later that Taib's nephew, a local politician, and two men who lived in the longhouse owned the company that had originally been granted the concession. They had sold it to WTK and pocketed a huge profit without doing any work at all.

Later on, WTK agreed to pay compensation of two ringgits (roughly 60 US cents) to Long Keseh for every tonne of timber taken from the communal forest—a tiny sum in comparison to the hundred US dollars or more that the timber company would be able to earn out of each tonne of timber.

BLOWPIPES AGAINST BULLDOZERS

In contrast to the majority of the inhabitants of his village, Harrison Ngau refused to be taken in. He has remained steadfast in his opposition to the logging companies. That made him extremely unpopular with the timber barons, but he was not dependent on WTK. Having finished school, he went on to work in a hotel in Miri, then in an ice factory, and after that in prospecting for crude oil under contract with Shell.

In 1980, Harrison took on the task of developing a Sarawak branch for the environmental and human rights organisation Sahabat Alam Malaysia (SAM), the Malaysian section of Friends of the Earth. "By the time Rahman Ya'kub's period in office as chief minister was reaching the end, clearing the rainforest had really become big business in Sarawak. Rahman's friends were granted big logging concessions and determined the political agenda from behind the scenes. Timber was fetching a high price, and trees were plentiful. At that time, no one had yet penetrated the forests in the interior of the country."

Harrison moved to Marudi, a rural commercial and administrative centre on the Baram River, rented a room there and made it into his office. In November 1980, he married 17-year-old Uding, a Kayan from a village near his own. They have raised four children while they continued campaigning against the ever-encroaching logging companies.

"In those days, environmental and human rights organisations were still completely unheard of in Sarawak," Harrison added. "The first organised protest in a loggers' camp was staged in the region of the Apoh River. That was early in 1980."[16] The intense protest was targeted against the Samling timber group. "Kayan from three longhouses entered the loggers' camp in large groups and threatened the Samling employees. The police called in by Samling were themselves indigenous Sarawakians and did not intervene to stop the Kayan. I supported the action in the background." In the end, Samling agreed to pay a commission to a cooperative comprised of the three longhouses.

"At that time, there were different views inside the communities," Harrison explained. "There were some people who wanted no logging at all. Others said: We're willing to accept it under certain conditions; if our communities get schools, roads and hospitals in exchange. Our position was: We support the indigenous communities in exercising their rights. What they want to do with those rights and what they want to demand is their affair. My job primarily involved advising the people, acting as a go-between and organising meetings between the communities as well as workshops."

By the mid-1980s, the Penan communities further upstream were more and more affected by the logging. "Every week, Penan delegations called in on our office. The Penan differed from the other communities in that they totally rejected any logging, which led to tensions between them and the Kayan, Kenyah, and Kelabit. Bruno Manser was already living with the Penan at that time and helped them to organise themselves. I did not meet him in person until much later when I travelled to Switzerland."

The big showdown came in March 1987, when thousands of indigenous people protested simultaneously against deforestation. Some 4,700 participants from 26 Penan villages and six longhouses erected blockades on logging roads to stop the loggers from penetrating any farther into the upper reaches of the Baram and the Limbang. Around 200 bulldozers and 1,600 timber workers were held up for several months. Penan armed with blowpipes stood in the way of the bulldozers, but remained completely peaceful at all times and refrained from any form of violence.[17] In a dramatic appeal to the government, they called on Taib to rescind the logging concessions and to call a halt to deforestation in the Sarawak interior.

Taib had no intention of acceding to such demands. Instead, he waited long enough for the protests to have lost much of their momentum, and for the indigenous people to be forced to return to their livelihoods. After a few months, only a few men, women and children were still tending the blockades. That was when he let loose the police to break up the blockades. That action was taken in liaison with the Malaysian federal government, which arranged for the arrest of more than a hundred of its opponents throughout the country in a coordinated action on 27 October 1987, without obtaining any court order. With this so-called "Operation Lalang" ("weeding-out operation"), the new strong man in Kuala Lumpur, Prime Minister Mahathir Mohamad, wanted to demonstrate what would be tolerated and what not.[18]

Harrison Ngau was working with his team in the office of Sahabat Alam Malaysia (SAM) when the telephone rang shortly after lunch. "The head of police in Marudi was on the line. He said it was about a threatening letter we had received because we were helping the Penan. We had reported the matter and filed a criminal-law complaint. He asked me if I had any travel plans for the coming days. I said no and that I was going to be in Marudi the whole time."

A quarter of an hour later, there was a knock on the door and six police officers forced their way into the narrow office. "I realised instantly what was happening," said Harrison, looking back. "A friend had warned me that arrests were planned and had suggested that I disappear into the interior. That, however, is not what I wanted, because I'm not the sort of person that runs away in situations like that."

The head of the squad, Inspector Yusuf, introduced himself: "I've come to arrest you under Article 73 of the Internal Security Act." The police seized all of Harrison Ngau's documents and started to take them away. He insisted, however, on signing every single confiscated paper to make it more difficult for the police to fabricate documents that they might claim had originated in his office. Reluctantly, the police agreed and it was not until six in the evening that they finished in the SAM office.

Harrison was flown to Miri and spent the night in a cell there. The next morning, he was driven by off-road vehicle along a bumpy country road to the police headquarters in Kuching, 512 kilometres away. Harrison was held for sixty days and nights without any charge being brought against him. "I was questioned at least ten times, at all hours of the day and night," says Harrison, recalling his time in prison. "Sometimes they woke me in the middle of the night to question me. Then they put a blindfold over my eyes and drove me around for several hours in a police car. They do that to make you lose your sense of orientation and to break your resistance. I had, however, nothing to hide, and everything I had done had been legal. In the end, the officers agreed that I was in the right and said that they would have acted in the same way if they had been in my position."

Harrison Ngau's arrest brought a flutter of publicity to the campaign against deforestation in Sarawak. Environmentalists and human rights campaigners all over the world sent letters of protest to the Malaysian government and called for a boycott of tropical timbers from Malaysia. "Many people in Sarawak who had kept quiet up until now spoke out publicly in my favour. That is when Taib and Mahathir realised that my arrest had had completely the opposite effect to what they had hoped for." Harrison was released at the end of December 1987. One year later, Sahabat Alam Malaysia received the Right Livelihood Award—considered the "alternative Nobel prize"—for Harrison's work in Sarawak. In 1990, Harrison was elected to the Malaysian federal parliament to represent the Baram constituency.

BRUNO MANSER, LAKEI PENAN

Four years after Harrison had moved to Marudi, a young Swiss man arrived in the region. The 30-year-old Bruno Manser had set out from Basel, Switzerland, for Borneo in 1984 in order to make a childhood dream come true. He joined a caving expedition in the Mulu National Park in northern Sarawak, after which he intended to seek the last nomads in the virgin forests of Southeast Asia.

"If only I could one day travel to Sumatra, Borneo, and Africa and live like a caveman there in the deep, impenetrable jungle amidst gorillas, orang-utans and other animals!" Manser wrote in an essay written when he was still at school. In the same text, the young man—whose father earned his living in the chemical industry—also revealed his radical "back to nature" utopia: "As a man, I would like to raze to the ground all factories that are not vital. Instead of them, I would bring to life a large forest with clear water and numerous animals."[19]

Bruno Manser, the passionate nature lover and friend of humanity, believed he had found his destiny in a life without money. After breaking off his studies in medicine, he spent several summers as a dairy farmer and shepherd in the Swiss Alps. It was then that he decided to get to know a primitive people still living very "close to the origins of humankind". A book in the university library in Basel put him on to the Penan. Was it perhaps Charles Hose's *Natural Man* that Bruno Manser had had in his hands there? Hose's description of the Penan as "noble savages" would surely have struck the eager Rousseauist in Manser like a flash of lightning.

Bruno Manser found his primeval paradise with Along Sega's Penan group, which at the time was in the Adang region of the upper Limbang Valley. He spent six years living with the nomads, and learned from them how to survive in the jungle. Manser's mentor adopted the talented and inquisitive "orang putih" (white man) as his foster child. From Along Sega the Swiss newcomer learnt how to hunt with the blowpipe and process the sago palm. He donned a loincloth and spent months wandering through the forests with Along's extended family. Having fled from civilisation, Bruno Manser adapted so well to the Penan that the nomads gave him the nickname of "lakei penan" (Penan Man).

Two years after Bruno Manser had found his heaven on earth in the rainforest, however, the life of the Penan began to undergo radical change. The incursion of the logging industry with its bulldozers and chainsaws threatened to destroy the habitat of this peace-loving people. Taib's "development policy" had caught up with Manser, the refugee from civilisation, right in his remote refuge. When prospectors from the logging industry advanced into Penan territory and the bulldozers became an increasing threat to the habitat of the last nomads, Manser climbed into the crown of a giant tree, where he wrote the following in his diary: "And when I behold the unspoilt valleys of the Seridan River—right up to the green swathe of the mountain ridges, where hardly any human foot has ever trod, I cannot stop the tears from coming to my eyes. Nature—you are Truth—even without human intervention. (...) And my heart cries like a funeral song—does this paradise really have to die and make way for chainsaws and bulldozers?"[20]

At first, Bruno Manser hesitated, but finally decided to make the cause of the Penan and the protection of the virgin forests of Borneo his principal mission in life. With the support of friends both inside and outside Malaysia, he began smuggling journalists into Borneo. He achieved his big breakthrough in the media in October 1986, when the German magazine *Geo* published an extensive report on the Penan's resistance to deforestation. It was translated into several languages and reprinted many times. The reporters had managed to travel into Sarawak disguised as forest scientists.[21]

In the years that followed, Bruno Manser organised gatherings of logging opponents in the forest, drew up petitions and instructed the Penan how to defend their rights by setting up roadblocks. The big wave of blockades in 1987 also bore his hallmark. While the Penan were blocking the logging roads, Bruno Manser remained in the background, hiding in the forest. The Malaysian authorities nevertheless got wind of the role he had played and declared that he bore responsibility for coordinating the Penan protests. "The Malaysians' reproaches were certainly true in the beginning," confirms Roger Graf, who was one of of Manser's companions at the time. "I know how very much he strengthened the timid Penan in their resistance and told them what they ought to do in order to jolt public opinion."[22]

Before long, Taib and the Malaysian government declared Bruno Manser *persona non grata* and he became the country's most wanted person. Despite a bounty of 50,000 ringgits (about 15,000 US dollars) offered for anyone catching Manser, several attempts to arrest him failed. Sympathetic indigenous people kept hiding him. He twice managed to avoid being captured by the police and military by fleeing into the jungle.[23]

UNCLE RAHMAN PUT OUT TO GRAZE

Events in March 1987 showed clearly that Taib could have frozen the logging concessions at any time, as demanded by the Penan and Bruno Manser. It was not the protest by Sarawak's indigenous peoples but a stunning feud between Taib and his uncle Rahman.

The first public disagreement between the two had occurred earlier, in September 1983. At that time, Rahman was still governor of Sarawak, and at the opening ceremony for a new port in Bintulu, he complained bitterly about the federal government in Kuala Lumpur which had promised the town of Bintulu a new airport but had not kept its word. Taib, who was only able to govern because he enjoyed strong support from Kuala Lumpur, was very upset with his uncle. He left the festivities without uttering a word, and it is reported that he immediately tendered his resignation.[24]

However, as it turned out, Rahman rejected Taib's resignation and resigned himself instead. Only the insistence of the Malaysian King made him stick it out until the end of his term in office. In April 1985, Rahman cleared the way for Taib, who had a loyal follower appointed as governor and set about marginalising Rahman's men in the administration, primarily by abolishing the forestry ministry and transferring logging concessions to a new ministry of resources planning. In July 1985, he took the leadership of this ministry for himself.[25]

"Taib in person has signed all the logging concessions and plantation licences since then, just he alone and no one else," says Harrison Ngau. "Before then, the forest used to be common land for us, the indigenous inhabitants, more or less entrusted to us as a loan from God, and it was our duty to look after it. Taib took on the role of minister of agriculture and forestry himself and he knows how much power goes with that."

It was now not just a matter of issuing concessions, for Taib could just as easily withdraw them. Exploitation of the all of Sarawak—worth billions of dollars—depended on just one person, with no checks and balances, no transparency and no public accountability. In March 1987, at one stroke, he withdrew more than thirty logging concessions from

Uncle Rahman's followers under the pretext that they had not abided by the forestry laws.[26]

The background to this settling of accounts was a dispute over political power in Sarawak, which had come to a head in the so-called "Ming Court affair". Two years after resigning as governor, Rahman felt the urge to return to power and for that purpose he had recruited a strong alliance of disgruntled Iban politicians and Malays. They met up early in March 1987 in the Hotel Ming Court in Kuala Lumpur. Out of the forty-eight members of the Sarawak parliament, twenty-seven called for Taib's resignation. The offended ruler, much affronted by this power play, reacted immediately by calling a new election.

Once again, the politics of timber was entwined with the electoral calendar. By freezing the logging concessions, Taib halted the flow of donations into his uncle's election funds. When it came to reissuing the concessions afterwards, Taib was able to give them to his own allies in exchange for significant kickbacks, which, in turn, he used to cement his own political power. Taib made it unmistakably clear that he would immediately and severely reprimand political disloyalty. Only those who fought unconditionally on his side received a share in the lucrative pickings to be had from chopping down the rainforests.

Estimates of what the parties spent on the electoral campaigns in Sarawak in April 1987 ranged from 20 million Malaysian ringgits to over 100 million (US$ 6–30 million).[27] Thanks to electoral coffers filled to the brim from the logging business, combined with support from Prime Minister Mahathir Mohamad, Taib prevailed. His camp won a clear election victory, with twenty-eight seats in the new parliament going to the Barisan National Coalition led by his PBB party, compared to the twenty seats won by the "Maju" group allied to his uncle.[28]

While Taib's propaganda merchants were rejoicing over the election outcome as a "gentleman's victory",[29] the victor imposed political and economic excommunication on his defeated opponents, driving many of them to ruin.[30] Wreaking revenge on his uncle became one of Taib's principal obsessions in the years that followed. He missed no chance to chastise his uncle, whose talent and charisma he had envied throughout his life, and even arranged for Rahman, who had the reputation of living the life of a "bon viveur", to be spied on, making sure that his life became utter hell.

Taib was "aggrieved and hell-bent on taking revenge, and everyone has been particularly afraid of him since that time," reports one of Rahman's former golf partners. "Whenever Rahman appeared on the golf course in Miri, suddenly there was not a soul left to be seen there. Nobody wanted to be spotted by Taib's spies on the golf course at the same time as Rahman." Any business people who had had dealings with Rahman lost their public contracts and were shut out of lucrative deals. Politicians and civil servants suspected of supporting Taib's uncle were sidelined. "In those days, I used to play golf with Rahman frequently, but in the end I had to stop. My son asked me to; he was afraid of losing his government job."[31]

It was not until more than twenty years later, on Rahman's 80th birthday in January 2008, that public reconciliation occurred between Taib, who was still chief minister, and his uncle, who had long since ceased to represent any form of political threat. "Blood is thicker than water," said Rahman in a speech before more than a thousand guests at his birthday celebration in the Hilton Hotel in Kuching. He then went so far as to grovel publicly and to announce that Taib was someone "whom I have always loved."[32] Taib played along and refrained from speaking. The battle of the giants had long been decided in his favour.

Whether it was in wrangling with his uncle, or in the dispute with the Penan protesting against deforestation, Taib never yielded ground unless he was absolutely forced to do so. His blatant lust for power led him to fight with whatever means he possessed, but he had the knack of delaying as well—holding his opponents at bay until they displayed a weakness. Then, once the right moment arrived, he would react in an instant, let loose his forces, and strike. Taib owed his political survival over five decades to exactly this tactic. He knew that nothing came cheaper than words and assurances; empty promises and threats were all part of his standard repertoire.

Power slowly changed Taib's personality and physical appearance. When travelling in Sarawak, he no longer went barefoot but had himself chauffeured in a Rolls Royce whose number plates heralded "the Most Honourable Chief Minister of Sarawak". When appearing in public, behind his wooden smile his eyes betrayed inner unrest. His intense red lips formed an odd contrast to his increasingly pale skin, his grey goatee beard and the gold frame of his oversized spectacles. Like many other despots, Taib had become paranoid and superstitious. For many years, a Pakistani *bomoh* (witch doctor) held mysterious exorcisms in the basement of his residence and became one of his most trusted political advisors.[33] Since the colour yellow was said to protect him, Taib developed a predilection for wearing yellow clothes, and, on his right hand, a heavy ring inset with a yellow diamond the size of a hazelnut.

In dealing with the outside world, Taib has always been fond of presenting himself as a statesman, and has been at pains to hone his rhetoric. It is impossible to count how often Taib has said that he is concerned about protecting the environment, and that he respects human rights, and how often he has assured the Penan of the millions being spent on their development or promised them vast areas of forest. Taib has always known full well that after a number of years nobody would remember such promises or that he would always be able to find some pretext or other for not abiding by them. It is also impossible to count all the occasions Taib has promised to resign, only to later put off his resignation. Possible successors have been chosen, played against one another, and then dropped. Taib was second to none in his mastery of the rhetorical game of possibilities—with the aim of leaving everything just as it had been before.

One man who has experienced Taib's power machine is the Swiss forestry expert Jürgen Blaser. Blaser spent many years working for Intercooperation, a Swiss development agency, and for the World Bank, and travelled on official missions to Sarawak twenty times. During those visits, he met Taib in person on three occasions. At the time of writing, Blaser is a professor at the University of Applied Sciences in Berne and advises the Swiss government on questions concerning tropical forests and the climate. In his office adjacent to woodland in Zollikofen, a suburb of Berne, the 58-year-old described his visits to Malaysia. "The first time I met Taib was with a delegation of Swiss parliamentarians in 1993. We were received in the lounge of his private residence. Taib acted in a very statesmanlike and modest manner. "Yes, there are problems with Mr. Manser and with the Penan, who have got to face up to development," he said. "Taib appeared to be interested in hearing our view and was absolutely convincing in the act he put on," Blaser reported pensively. "The atmosphere only became tense at a press conference in the Kuching Hilton a few days later when local journalists became very aggressive towards the Swiss delegation and accused us of interfering in their country."[34]

Blaser had a similar experience to report from his second visit in 1997 when he travelled with a World Bank delegation to present a 20-million-dollar project for a biosphere reserve in the north of Sarawak to protect the Penan rainforests. Once again, Taib presented himself to the guests from abroad as a "good guy"; he had the delegation collected in Mercedes limousines from the airport with a police escort and received them like royalty. The US and Swiss ambassadors to Malaysia were among the members of the party. "I left that meeting in a positive frame of mind and said to myself: Taib is most definitely interested in this project and the 20 million. But I could not have been more wrong. I was bombarded with criticism the next day."

Taib left it up to his cabinet colleague, James Wong, to play the role of the "bad guy". Wong was minister of the environment at the time and also a logging entrepreneur with huge concessions on Penan territory. Wong waited until the ambassadors had left and then took Blaser to task. "Wong barked at me that I was up to no good. A subordinate brought out a map of Sarawak, on which he showed me the Rejang River, which runs across the middle of the country: 'You are not to put one foot beyond this river,' he said. Of course, the territory of the World Bank project was on the other side of the river. The principal reproach was that I had allegedly been sent by Bruno Manser, although I was part of the official World Bank delegation."

In the end, the World Bank was offered a very small project at the opposite end of Sarawak. And it has still not proved possible to create the biosphere reserve, despite that fact that international funding is in place.

It was not until after Manser's disappearance that Blaser was permitted to travel to the north of Sarawak again and to play his part in the implementation of the Pulong Tau National Park, which had no real effect on the logging industry.[35]

SARAWAK MONOPOLY

With his victory in 1987 over his uncle Rahman, Taib had consolidated his political power. His next goal was to bring Sarawak entirely under his economic control as well. The instrument he used to this end was the state-run Cahya Mata Sarawak (CMS), the largest business in East Malaysia, whose main product in those days was cement.

By means of a carefully arranged series of privatisations, spin-offs and newly created companies, CMS managed to grow into more lucrative areas of state business between 1993 and 1996, before the Taib family became the company's new proprietors.

"Reverse takeover" is the term applied to the complicated process whereby CMS bought Taib family companies at inflated prices and paid for by means of a share swap, leaving the Taibs suddenly as majority proprietors of the state enterprise.[36] A simpler term for the whole exercise would be fraud. In this way, the closest members of Taib's family, namely his wife, Laila, and his four children, became the principal shareholders of the largest private company in Sarawak, which was kept afloat by public contracts running into millions of dollars.

With the acquisition of the majority holding in CMS, the Taib family, which was already in control of timber export, suddenly gained four further lucrative economic monopolies in Sarawak: production of cement; production of steel; trade in equities; and the Islamic bank and CMS subsidiary, the Utama Banking Group (UBG).[37] It goes without saying that setting up this monopoly had only become possible thanks to never-ending political manipulations.

One example is the finance company Sarawak Securities. It was taken over by CMS shortly after its establishment in 1992, and, despite having no business experience, was granted one of the highly coveted and difficult-to-acquire equity trading licences by the Malaysian minister of finance.[38] A further example is the 1993 acquisition by CMS of highly profitable companies in mining and steel production. The seller? The state of Sarawak.[39]

In only a few years, the state cement producer CMS had become a broadly diversified conglomerate with activities in securities trading, road building and maintenance, water supplies, the exploitation of quarries, the manufacture of cement, the production of steel and wire, and trading and investment.[40] Andrew Aeria, a professor of political science at the University of Kuching, who studied the history of CMS, arrived at the following conclusion in his dissertation: "This consolidation made CMS the unrivalled infrastructure and financial conglomerate in Sarawak and one of the major players on the Kuala Lumpur Stock Exchange."[41]

That the Taib family had now acquired control over CMS did not go unnoticed by the public in Sarawak either, where everyone knew everyone else in the relatively small business community. In 1995, Taib's sons, Abu Bekir and Sulaiman, became directors of the company, and Taib's brother Onn assumed the role of chairman. Another long-standing acquaintance of Taib, the architect Hijjas Kasturi, a fellow student in Adelaide in the days of the Colombo Plan, also made his way onto the CMS board. In the years since Adelaide, he had set up in business in Kuala Lumpur and had also become more or less court architect to the Taib family (see chapter 4).[42]

This gave rise to a new interpretation of the initials "CMS". Instead of Cahya Mata Sarawak (the "Light of Sarawak's Eyes"), people began to say "Chief Minister and Sons" or even "The Con Man of Sarawak".[43]

WAR OVER THE RAINFOREST

Despite all the police actions, arrests and threats, the Penan resistance to the clearance of virgin forests continued even after the major blockades of 1987. A new wave of blockades was staged in autumn 1989, and some

4,000 indigenous forest dwellers joined in. A hundred and seventeen Penan were arrested and taken into custody.[44] Around that time, Bruno Manser, who was suffering from a snake bite, increasingly gained the impression that he would be able to help his indigenous friends more by acting from abroad. At the beginning of 1990, after spending six years in Sarawak, he returned to Switzerland, travelling under a false name, and, once there, set up the Bruno Manser Fund (BMF).

A world tour by a Penan delegation, protest actions in Japan in front of the headquarters of the Marubeni group (one of the principal purchasers of timber from Sarawak), and a sixty-day hunger strike in front of the Swiss federal parliament building in Berne were amongst Manser's most important actions in the early 1990s. In between, he kept returning covertly to his partners in Sarawak, crossing the land border—despite an entry ban placed on him by the Taib government.

One of Manser's most important partners in Sarawak was Mutang Urud—author of the foreword to this book—who was a politically canny Kelabit from Long Napir, a village in the upper reaches of the Limbang River, on territory included in the logging concession held by the minister of the environment, James Wong. Mutang arranged secret visits to the Penan by foreign journalists, organised blockades, and coordinated the resistance of the indigenous communities to deforestation. At the end of 1990, he accompanied two Penan representatives and Bruno Manser on a seven-week tour of twenty-five cities in thirteen countries around the globe to disseminate information about the situation in Sarawak. [45]

Considerations of safety and security had led the Sarawak Indigenous Peoples' Alliance (SIPA) set up by Mutang to move its head office to the island of Labuan, just off the coast of Borneo. From there it was only a short journey by speedboat to the coastal towns of Limbang and Kota Kinabalu as well as to Brunei.

Mutang, however, had to go through the same experience as Harrison Ngau before him. The Malaysian security machinery, in particular the dreaded Special Branch—the political police—would tolerate only so much. In Mutang's case, it happened on 5 February 1992, at a time when the security forces were trying to clear a large Penan roadblock near Long

Ajeng in the Upper Baram. Since mid-May 1991, more than 500 Penan had been preventing the Samling logging company from accessing the Upper Baram.

Back in the 1980s, Mutang had already been compelled by a relative of his, a high-ranking police officer, to report Bruno Manser's whereabouts regularly to the Special Branch. This time, the anger of the security apparatus was directed against him. An apparently friendly neighbour turned out to be a police spy, and the officers marched into the SIPA office and arrested him, under the pretext that his organisation had not been officially registered. It was while he was being held in custody that the Long Ajeng blockade was cleared by the Federal Reserve Unit, a police unit specialising in brutal suppression of public protests.

In violation of the freedom of association—a fundamental human right protected by international treaties—all unregistered clubs and organisations with more than seven members were considered illegal in Malaysia. The strict monitoring and supervision of all organisations by the Registrar of Societies (ROS), an arm of the Ministry of Home Affairs, is one of the Malaysian government's most effective tools for keeping political control over society. Applying the euphemisms of official jargon, the ROS has described its objectives as ensuring "the growth and development of a healthy and orderly society, which is not in conflict with the requirements of peace, welfare, security, public order or morals."[46]

The Penan resistance to forest clearance was most definitely viewed by Taib and his friend Mahathir in Kuala Lumpur as anything but a "healthy and orderly society", and so Mutang Urud had to go to prison. When he was released on bail on 3 March 1992 under a court order, after a month in solitary confinement, his enemies did not delay. Less than 45 minutes after his release, Mutang was again arrested by the police, this time invoking an emergency ordinance.[47]

Prime Minister Mahathir Mohamad, a former medical doctor who was little enamoured with democracy and the rule of law, and who was fond of haranguing the west, wrote a defamatory letter to Bruno Manser in Switzerland that very same day. He threatened to hold Manser personally responsible for anyone injured in the course of the combat against

Penan resistance, though that resistance itself always remained non-violent. To quote the letter: "If any Penan or policeman gets killed or wounded in the course of restoring law and order in Sarawak, you will have to take the blame. It is you and your kind who incited the Penans to take the law into their own hands and to use poison darts, bows, arrows and parangs to fight against the Government. (...) Stop being arrogant and thinking that it is the white man's burden to decide the fate of the peoples of this world. (...) You are no better than the Penans."[48]

Following a massive international campaign for his release, Mutang Urud was allowed to go free in April 1992. The harsh prison conditions, however, had left their mark on him; he decided to emigrate and went to live in Canada. In May 1992, Mutang accompanied Bruno Manser to the UN Earth Summit in Rio de Janeiro. In December of the same year, he addressed the UN General Assembly in New York.[49] After all this, Mutang went on to study anthropology in Vancouver, marry a Canadian and raise a family. Twenty years elapsed before he saw his homeland again.

THE PENAN'S NEW PROTEST CULTURE

However, the crushing of the Sarawak Indigenous Peoples' Association (SIPA) did not break the Penan resistance. The brutal response to their blockades had only encouraged them; now they knew how to stand up to the loggers. While increasing numbers of indigenous peoples submitted, the apparently docile "noble savages" displayed the utmost determination in defending their traditional territories. After Bruno Manser and Mutang Urud had left, they had become more self-reliant than ever. In the meantime, a new generation of Penan leaders was ready to take over the coordination of the protests.

A year after the police action near Long Ajeng, the Penan once again set up a blockade in the Upper Baram, on the edge of their traditional territory.[50] A blockade village—even including a church—was to assure that the protest site could be permanently occupied. The blockade held for six months until 300 police officers and employees of the forest department smashed it with teargas and bulldozers on 28 September 1993. Eleven Penan were arrested, and a 4-year-old boy died, presumably of teargas poisoning.[51] He was not the only victim of the conflict between the Penan and the timber industry. On 8 September 1994, a 60-year-old Penan died after extensive use of teargas by the police in an action against a blockade near Long Mubui.[52] Later Abung Ipui, a protestant minister and block-ade leader from the community of Ba Kerameu, was murdered in October 1994—presumably by loggers. The Penan found him with his stomach slit open, floating in a river near his home village.

One Penan elder recalls that "It was like war at that time. Samling and other logging companies were absolutely determined to make inroads into our forest so that they could chop down the valuable trees there. In doing that, they were supported by the Taib government and the police. No one was interested in our rights."

That was the time when Joe Jengau Mela, a Penan from the Upper Baram, became one of the leading personalities in the resistance to the loggers. After attending a primary school run by Australian missionaries belonging to the Borneo Evangelical Mission, he successfully completed the state secondary school in the 1970s. Like many young Penan, his first job on leaving school was as a guard for a timber company. He had to make sure that no logs were stolen from a timber depot. When many people from his home village were arrested during the wave of blockades in the early 1990s, the devout Christian visited them in prison. That led him to end his employment with the timber company and turn instead to low-paying jobs at a filling station and as a cook in the town of Miri. It was not long before he became an important contact point for Penan visitors to the town, providing a link with Bruno Manser and NGOs abroad, first by telephone and fax and later by email. After he had trained as a commu-nity organiser in Kuching, the Penan headmen appointed him to be their external representative. It was Joe Jengau Mela who, in 1998, along with the headmen of four Penan communities from the Upper Baram, went to consult the indigenous lawyer, Harrison Ngau, in order to file the Penan's first land rights claim in the territory that was later to become the Penan Peace Park.

Since times immemorial, the Penan have been roaming the rainforests of Borneo as nomadic hunters and gatherers.

Images from the Penan Peace Park in Sarawak's
Upper Baram region, documented in 2012 by French
photographer Julien Coquentin.

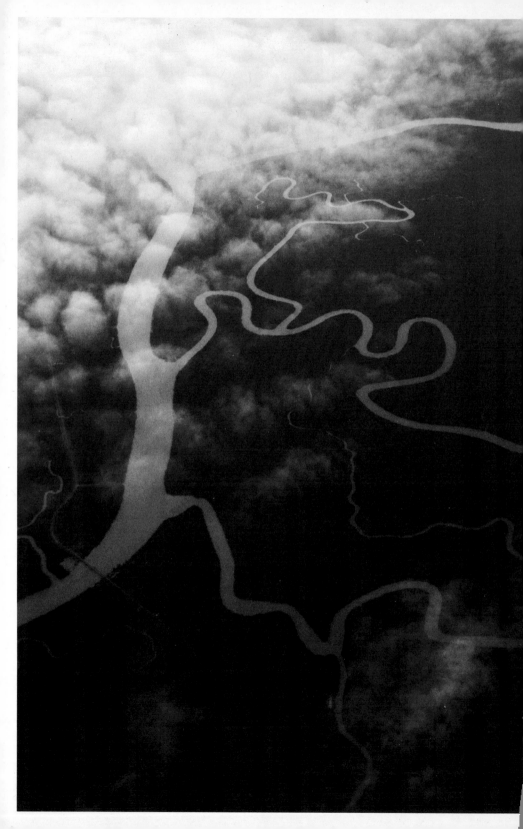

SARAWAK LOG PRODUCTION 1975–2013
(LOG PRODUCTION IN M^3)

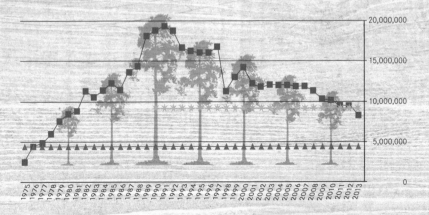

- ■ Log production in m^3
- ▲ FAO recommendation in m^3
- ✳ ITTO recommendation in m^3

SARAWAK'S SIX LARGEST TIMBER GROUPS
AND THEIR BUSINESSES 2010

Company	Timber concessions in Sarawak in ha	Plantation concessions in Sarawak in ha	International timber business	Shipping	Trade	Construction	Natural resources	Property development	Media	Finance	Tourism
Samling Group	>1,300,000	>632,876	x	x	x	x	x	x			x
Rimbunan Hijau	>1,000,000	685,073	x	x	x	x	x	x	x	x	x
WTK Group	850,000	264,472	x	x	x	x					x
Ta Ann Group	>577,000	413,644	x					x			
KTS Group	N.A.	430,909	x	x	x	x	x	x			x
Shin Yang Group	N.A.	372,918		x	x	x	x	x			x

Sources: 1 STIDC; Forest Department Sarawak; ITTO; Jomo et al. 2004.
        2 Annual Reports; KLSE; Stock Market Announcements; Faeh 2011.

The rainforest of Borneo is one of the most biodiverse habitats in the world. The rhinoceros hornbill (above) and the Sunda clouded leopard (below, left) are only two of thousands of animal species. Plant life in Sarawak includes dozens of species of wild ginger (above, right) and a fascinating variety of tropical fruit.

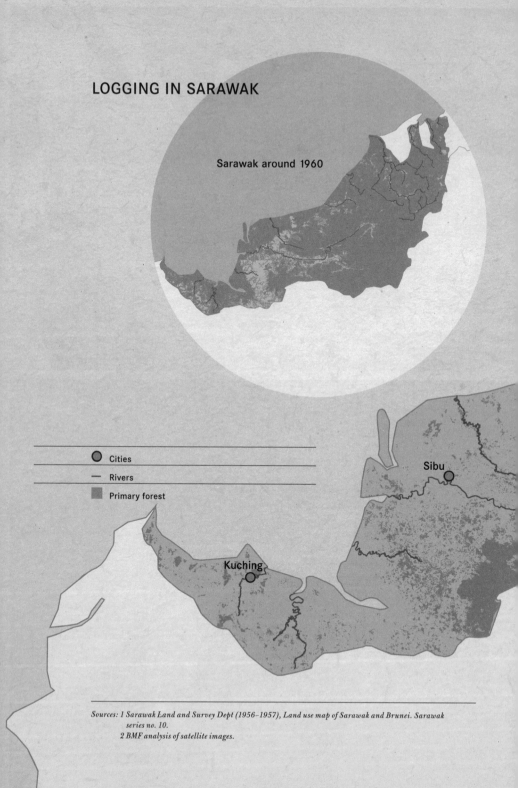

LOGGING IN SARAWAK

Sarawak around 1960

Cities

Rivers

Primary forest

Sibu

Kuching

Sources: 1 Sarawak Land and Survey Dept (1956–1957), Land use map of Sarawak and Brunei. Sarawak
series no. 10.
2 BMF analysis of satellite images.

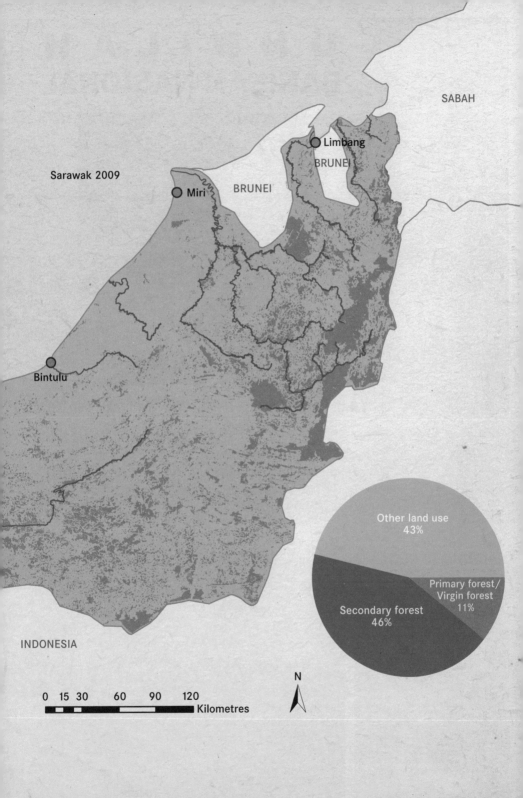

Sarawak 2009

SABAH

Limbang
BRUNEI

Miri
BRUNEI

Bintulu

INDONESIA

Other land use
43%

Primary forest/
Virgin forest
11%

Secondary forest
46%

0 15 30 60 90 120
Kilometres

N

U N D I L A H
BARISAN NASIONAL

Y.A.B. Datuk Patinggi Tan Sri (Dr.) Haji Abdul Taib Mahmud

Starting in the late 1980s, Swiss rainforest advocate Bruno Manser (above, right) and his mentor, nomadic Penan headman Along Sega (right, below), inspired hundreds of Penan to challenge the Taib regime's logging policies. Their peaceful blockades of logging roads (below) caught the world's attention.

RECONSTRUCTION OF THE FLOW OF ILLICIT FUNDS FROM SARAWAK'S TROPICAL TIMBER EXPORTS

TAIB
CHIEF MINISTER

SARAWAK MONOPOL

6

MALAYSIA

DEWAN NIAGA

ONN MAHMUD

1

2

1   Establishment of a monopoly for the export
      of tropical timber

2   Granting of timber export permits to Japan

3   The timber importers pay kickbacks to a shell
      company in Hong Kong

TIMBER
IMPORTERS

JAPAN

3

$

6 →

SAKTO GROUP
- - - - - - - - - - - -
JAMILAH TAIB

CANADA

4   The shell company transfers the kickbacks to
    an investment firm in Hong Kong
5   The investment company grants a loan for
    the purchase of real estate in Canada
6   Secret control over the company

5

**RICHFOLD
INVESTMENT**
- - - - - - - - - - - -
ONN MAHMUD

HONG KONG

4

REGENT STAR
- - - - - - - - - - - -
SHEA KIN KWOK

Source: BMF 2013

Since the early 1980s, hardwood timber worth an estimated 50 billion US dollars has been logged in Sarawak. Once the forests have been logged, they are prone to erosion and further degradation.

THE WORLD'S TEN LARGEST TROPICAL TIMBER IMPORTERS AND EXPORTERS IN 1991 (TIMBER VOLUME IN M³)

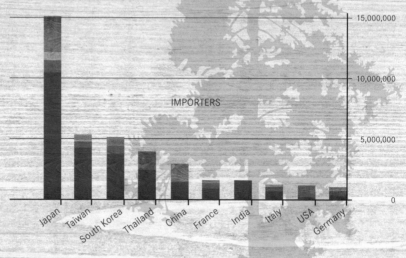

LOGS
VENEER
SAWN WOOD
PLYWOOD

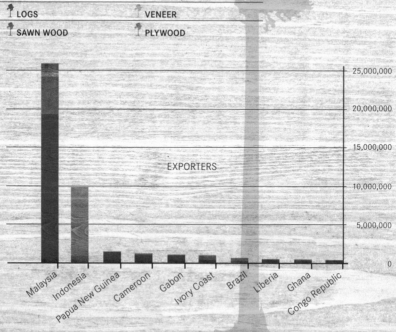

Source: ITTO Annual Report 1992

THE WORLD'S TEN LARGEST TROPICAL TIMBER IMPORTERS AND EXPORTERS IN 2011 (TIMBER VOLUME IN M³)

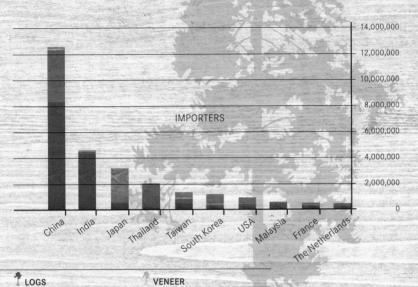

IMPORTERS

14,000,000

12,000,000

10,000,000

8,000,000

6,000,000

4,000,000

2,000,000

0

China India Japan Thailand Taiwan South Korea USA Malaysia France The Netherlands

🌴 LOGS   🌴 VENEER

🌴 SAWN WOOD   🌴 PLYWOOD

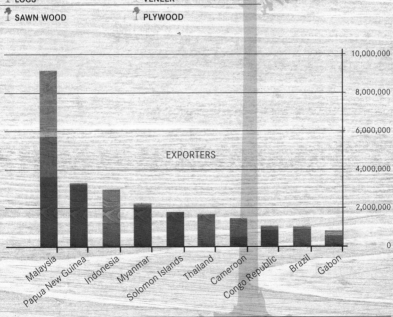

EXPORTERS

10,000,000

8,000,000

6,000,000

4,000,000

2,000,000

0

Malaysia Papua New Guinea Indonesia Myanmar Solomon Islands Thailand Cameroon Congo Republic Brazil Gabon

Source: ITTO Annual Report 2012

WORLDWIDE LOGGING BY SARAWAK TIMBER GROUPS SINCE 1990

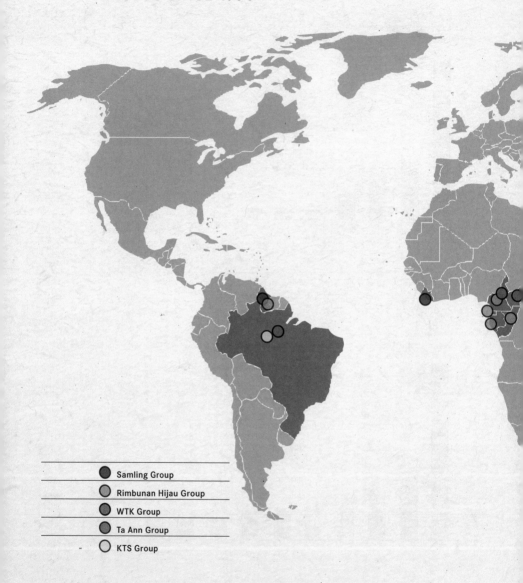

- Samling Group
- Rimbunan Hijau Group
- WTK Group
- Ta Ann Group
- KTS Group

Sources: Annual Reports; Faeh 2011; www.forestsmonitor.org; Global Witness

The Taib regime's plans to flood hundreds of square miles of tropical rainforest and forcibly displace thousands of Borneo natives have triggered fierce protests from Sarawak's indigenous communities.

DAM PROJECTS IN SARAWAK (2014)

Dam project	Status	Area (km²)	Affected villages	Water level (MAMSL)	Capacity (MW)
Bakun	completed	700	31	255	2400
Baleh	planned	527.3	1	241	1300
Baram 1	planned	412.5	36	200	1200
Baram 3	planned	6.3	4	435	300
Batang Ai	completed	76.9	59	125	108
Belaga	planned	37.5	0	170	260
Belepeh	planned	71.8	5	570	114
Lawas	planned	12.4	1	225	87
Limbang 1	planned	6.3	1	100	42
Limbang 2	planned	41.3	11	230	245
Linau	planned	52	3	450	297
Murum	under construction	241.7	10	560	944
Pelagus	planned	150.8	78	60	410
Trusan	planned	47.4	5	510	200

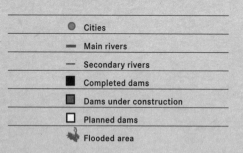

- ● Cities
- ▬ Main rivers
- — Secondary rivers
- ■ Completed dams
- ▨ Dams under construction
- ☐ Planned dams
- 🦢 Flooded area

Sources: Sarawak Energy; BMF 2013

Mukah

Sibu

Batang Ai

MALAYSIA
INDONESIA

SOUTH CHINA SEA

Lawas

Limbang

BRUNEI

Trusan

Miri

BRUNEI

Limbang 1

Limbang 2

Baram 1

Bintulu

Baram 3

Belepeh

Belaga   Bakun

Murum

Linau

Pelagus

Baleh

INDONESIA

N

0   25   50
Kilometres

GLOBAL PALM OIL CONSUMPTION IN MILLION METRIC TONNES 1992–2014

60
50
40
30
20
10
0

1992/93 1993/94 1994/95 1995/96 1996/97 1997/98 1998/99 1999/00 2000/01 2001/02 2002/03 2003/04 2004/05 2005/06 2006/07 2007/08 2008/09 2009/10 2010/11 2011/12 2012/13 2013/14

■ Palm oil   ■ Palm kernel oil

GLOBAL CONSUMPTION OF VEGETABLE OILS
IN MILLION METRIC TONNES 1992–2014

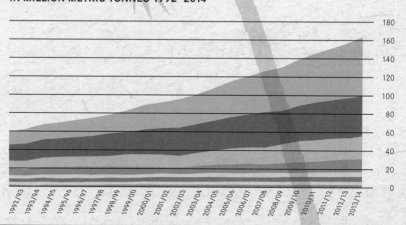

180
160
140
120
100
80
60
40
20
0

1992/93 1993/94 1994/95 1995/96 1996/97 1997/98 1998/99 1999/00 2000/01 2001/02 2002/03 2003/04 2004/05 2005/06 2006/07 2007/08 2008/09 2009/10 2010/11 2011/12 2012/13 2013/14

■ Palm oil/Palm kernel oil   ■ Sunflower oil   ■ Coconut oil
■ Soybean oil   ■ Cottonseed oil   ■ Olive oil
■ Rapeseed oil   ■ Peanut oil

Source: USDA statistics

The replacement of rainforest by oil palm plantations alters the landscape permanently.

Many indigenous communities have lost their lands to oil palm companies without

being consulted.

Stop timber corruption in Sarawak!

Freeze Taib assets n

er-corruption.org

On the occasion of Taib's 30th anniversary as chief minister of Sarawak in 2011, rainforest campaigners protested in front of Taib family properties in Ottawa (1–2), Seattle (3–6) and London (12). More protests were held at the Malaysian Embassy and Parliament House in Berne, Switzerland (7–9); and at the Hobart Harbour (10) and at the University of Adelaide, Australia (11).

In one hectare of tropical
rainforest, more tree species
can be found than in the
whole of Europe. If future generations are to know the
natural treasures of the Borneo
rainforest, the determined
support of the international
community is needed.

In the course of the 1990s, developments among the Penan went two different ways. Though the well-organised Penan communities in the Upper Baram—who were determined to offer resistance—succeeded in protecting considerable parts of their virgin forest against the chainsaws, the small groups of nomads in the Limbang region were not able to stop the incursion of logging companies in their territory. Manser could do no more than look on as the paradise of the last hunter-gatherers, which he had discovered only a few years before and meticulously described in his diaries, was being destroyed ever faster, year after year.

MANSER'S LAST JOURNEY

The nomads' lack of success in the virgin forest was heart-breaking for Bruno Manser, and he began to take ever greater risks in his protest actions. He went on record publicly as saying that he was pushing the boundaries of safety in the actions he was taking; increasingly frequently he crossed those boundaries too. In a daredevil action in the Swiss Alps in 1996, he attached himself to the cable of the Klein Matterhorn cable car and slid down 800 metres to congratulate the city of Zermatt on entering into a commitment to make no further use of tropical timber. In April 1998, he made a parachute jump over the United Nations in Geneva in order to offer the Malaysian UN embassy a lamb as a token of reconciliation on the occasion of the Islamic festival of sacrifice.[53] One year later, he steered his motorised hang glider to land in front of Taib's residence in Kuching.

Spectacular as Manser's actions may have been, for his fellow campaigners they were an expression of the environmentalist's lack of a definite plan following Taib's veto of a planned biosphere reserve. Roger Graf, for instance, who was the secretary of the Bruno Manser Fund then and is head of information and education at Zurich Zoo at the time of writing, says: "I felt a growing feeling of dissatisfaction, since I noticed that we were simply marking time. It often happened that Bruno came into the office after meeting a prominent figure and was convinced that things would now start to move. He always used to believe in the good in all people. It was enough to make you despair."[54]

Given the rapid destruction of the forest, and the apparent ineffectiveness of his actions, the only solution Manser saw was to seek reconciliation with Taib. He wrote flattering letters to him and offered to make peace. Manser, who was fond of quoting the Chinese philosopher Lao-tzu, was unshakably convinced that it was in the nature of all people to be good. It is quite clear that Manser had picked the wrong man in Taib to test his naïve belief in the universal good of human beings. The chief minister remained totally immune to the advances of the Swiss troublemaker; he never replied to any of Manser's letters.

"I am tired, but I cannot stop until the Malaysians' promises have been honoured—the biosphere reserve and self-determination for the Penan promised in 1987,"[55] is how Manser himself put it in 1998 when speaking to the Basel-based journalist, Ruedi Suter, who was later to become his biographer. One year later, he summed up a frustrating decade of total commitment with the words: "Success in Sarawak is less than zero."[56]

At the end of February 2000, Manser travelled to Borneo again. He spent three months in Kalimantan, the Indonesian part of the island, working with the BMF secretary, John Künzli, and a Swedish film team on the completion of a film, started ten years previously, about his rainforest campaign. After that, it was Manser's intention to visit his Penan friends in Sarawak. Arriving on foot from Indonesia, Bruno Manser crossed the Malaysian border incognito, not far from Bario, deep in the Borneo interior, on 23 May 2000. The day after that, he set out from Bario, accompanied by a Penan friend. On 25 May, Manser took leave of his friend to continue several days' trek alone through the jungle to meet up with his friend and mentor, Along Sega, in the Adang region, passing the Batu Lawi mountain he so loved. From that day all trace of Bruno Manser has been lost. He was never seen again. His 25-kilo rucksack was never found, nor were any other clues as to his whereabouts. Five years after his disappearance, on 20 March 2005, the civil court of the canton of Basel-Stadt formally declared Bruno Manser to be missing, presumed dead.

Did Bruno Manser suffer an accident in the jungle, was he murdered, or did despair lead him to take his own life on his last journey? Despite extensive investigations and search expeditions, his fate remains unaccounted for. One thing, however, is beyond doubt: Manser, the adopted Penan, became the victim of his thwarted dream of saving the virgin forests of Sarawak.

BRUNO MANSER'S LEGACY

After the disappearance of its founder, the Bruno Manser Fund faced major challenges. The most pressing jobs were mapping the Penan rainforests and filing land rights claims in Sarawak. While this was happening, one of the principal claimants died in mysterious circumstances. Shortly afterwards, a Penan woman broke the silence on sexual abuse by logging company workers.

BREAKTHROUGH IN COURT

At the beginning of the 1990s, the Penan and other indigenous communities in Sarawak had already begun to map their traditional hunting and foraging territories in the rainforest with the support of Canadian and US NGOs. "We wanted to put a strong symbol in the hands of the Penan for their blockades," recalls the human rights activist Mutang Urud. "We were told that the maps from the colonial period had been lost in a fire at the district office in Marudi. Our lawyers, nonetheless, insisted that they needed maps urgently to document the legitimacy of the indigenous claims to land rights."

When the Canadian First Nations communities achieved a groundbreaking victory for their indigenous land rights in 1997,[1] it inspired the indigenous peoples of Sarawak to do the same. A central role in Malaysia was played by Baru Bian, a minister's son from the people of the Lun Bawang, and the first indigenous lawyer to practise in Sarawak. Baru Bian was born in 1958 and studied law in Peninsular Malaysia before going on to complete his studies in Melbourne. He then returned to Sarawak, and in 1992 set up his own legal practice in Kuching under the name of Messrs Baru Bian Advocates & Solicitors. He found a partner in See Chee How, a young lawyer of Chinese origin and a human rights activist from Kuching. It was not long before the two of them were involved in numerous land rights cases, from which other legal specialists had shied away in fear of reprisals from the Sarawak government and the powerful logging companies. Baru and See's work would lead them to enter politics two decades later, when the two of them were elected in 2011 to the Sarawak state assembly (regional parliament) as members of the opposition People's Justice Party (PKR).

The two lawyers achieved a spectacular breakthrough in May 2001 with the so-called Rumah Nor case. It had been brought by Headman Nor anak Nyaway of an Iban community against the plantation group Borneo Pulp Plantation and the Taib government. Without consulting the Iban, the government had issued a lease on large parts of the Iban's communal rainforest for pulpwood plantations.[2] The victory for the Iban was the first time a Malaysian court recognized indigenous land rights over primary rainforest. The courts ruled that the intact virgin forest was a central element of the communal land, used for hunting and gathering of forest products. "I'm really overjoyed at this important precedent on which we have been working for a long time," said a jubilant See Chee How in his comments after the court's decision.

The judgment, handed down by Judge Ian Chin of the Sarawak High Court, demonstrated astonishing independence from the Malaysian government. Chin knew the price of that independence. After a much-maligned judgment against a politician belonging to the ruling Barisan National government in 1998, he had been verbally threatened by Prime Minister Mahathir Mohamad, and then enrolled in a five-day boot camp with other judges for "re-educational" purposes. While there, the primacy of the government's interests was hammered into the judicial civil servants.[3] Crushing the independence of the courts was done systematically under Mahathir. In 1988, the autocratic Premier had arbitrarily dismissed the country's top judge, Lord President Salleh Abas, thereby keeping the remaining judges on a short lead.[4] Even today, in 2014, Malaysia's judges still have difficulty ruling independently when government interests are at stake.

Taib was livid at the court ruling in favour of the Iban and ordered J. C. Fong, the attorney general—one of his closest political allies—, to appeal without delay. In order to pre-empt a flood of land rights claims from indigenous people, the Taib government arranged to have a new bill immediately placed before the state assembly. This "Land Surveyors Bill" outlawed the surveying of land by anyone other than state surveyors and was intended to prevent indigenous peoples, under threat of criminal sanctions, from drawing up maps of their land, which they would be able to use in land rights claims.[5] This constituted a clear violation of the elementary human rights now codified in the United Nations Declaration on the Rights of Indigenous Peoples.[6]

Though the Land Surveyors Act made community mapping an illegal activity, Taib's new law could not prohibit the courts from accepting such maps. The Rumah Nor judgment breached the dam. Nothing could halt the flood of new indigenous land rights claims. A 2010 report found that more than 140 communities had already filed collective claims.[7] Since then, the number of cases has risen to over 200.

POWER OF HISTORY

Headman Nor's key evidence was proof that indigenous people had been occupying their claimed native lands before 1958—and that they had used it continuously since then. In the British colonial period, a new Sarawak Land Code had come into force, placing limits on the traditional use of land by the indigenous peoples. After 1958, indigenous communities needed state approval to move to new territories, and the practice of granting approvals was extremely restrictive. One of the intentions was to protect the Sarawak forest as state land from any occupation and reserve it for government use. In contrast to the customary law of the indigenous peoples of Borneo, which was known as *adat*, the British Crown only recognised farmed land as "native customary land".

For the rainforest culture of the Penan, with no tradition of the written word, producing documentary evidence of historic land use represents a particular challenge. It can only be met by systematically recording oral history and extensive cartographical work in the forest. The Penan researcher Joe Jengau Mela spent many years documenting the legends and history of the Penan Selungo, who continued to live as nomads into the second half of the 20th century. They were the people whom Rodney Needham had described as "Eastern Penan", and their traditional territory was on the Selungo River.

The Penan's own oral tradition was supported by the first-hand accounts of much more recent times by the government officers, anthropologists, and missionaries who visited them. One of these was Ian Urquhart, a former British colonial officer, whom I met with his wife, Bunty, at their house in South Croydon, not far from London. The two of them gave me an enthusiastic account of the more than eighteen years they had spent together in Sarawak—the best time of their life, they said. To prove his affection, the 86-year-old rolled back a sleeve and proudly showed me the large tattoos indigenous Iban had marked on his upper arm.

Ian Urquhart was born in 1919. He fought for the British in India in the Second World War, and then took a position in the British colonial service. He set foot in Sarawak in 1947 as a young officer and was one of the first three colonial officers in the service of the British Crown who replaced the Brookes's long-serving officers.[8] Urquhart had been stationed in Marudi from 1955 to 1957 as the district officer for the Baram, and had carried out detailed linguistic and anthropological studies, which he published in the *Sarawak Museum Journal.*[9]

Urquhart undertook eight expeditions to the territory near the source of the Baram River and its tributaries, which was the traditional land of the indigenous peoples of the Penan, Kenyah, Kayan, and Kelabit, deep inside Borneo. An enthusiastic amateur photographer, he made 8-mm colour films of his boat negotiating the rapids in the Akah and Baram rivers, of a missionary aircraft from the Borneo Evangelical Mission dropping food, and of the Penan nomads performing dances to welcome him—documents of inestimable historical value.

"The Penan nomads were the only inhabitants of Sarawak who did not have to pay any taxes," Urquhart recalls, "which explains why we never collected much information about them. All the other indigenous peoples had to hand one Sarawak dollar per family per year over to the state. Whenever we went out into the field, we took lists containing the names of all longhouse inhabitants with us in waterproof tins."

A trek into the rainforest in 1957 took the Urquharts to the territory around the source of the Selungo, where they were accompanied by Penan. "The Penan led my wife and me across country and carried our luggage for us," Urquhart recalled. For him, it was clear that the Penan had been living in the rainforests of the Baram, Tutoh and Limbang basins for a very long time, and he expressed the hope that this land would be rightfully awarded to them.[10]

The oral history interview with Ian Urquhart, who died in 2012, is part of the documentation for a land rights case being pursued by the Penan. Four communities from the Upper Baram region joined forces in

1998 to stake a legal claim to 415 square kilometres of virgin forest and rice fields. The case was filed by the Headmen Kelesau Na'an (Long Kerong), Bilong Oyoi (Long Sait), Pelutan Tiun (Long Sepigen), and Jawa Nyipa (Long Ajeng). The 190 Penan families (approximately 800 individuals) living in these four villages claimed the land around the Gunung Murud Kecil, a mountain ridge at an altitude of 1,600 metres near the border with Indonesia.

The rainforest around the Gunung Murud Kecil is one of the only virgin forests remaining in Sarawak. Centuries-old trees are still standing thanks to the blockades of the 1990s and the vigorous resistance by the Penan of the Upper Baram against the chainsaws of the Samling logging group.

A PALM FOR BRUNO MANSER

The Penan headmen were waiting in a semicircle in a clearing in front of the village of Long Kerong. Straw hats with rhinoceros bird feathers, batik shirts and necklaces of colourful glass beads distinguished the elders of the seventeen communities in the Upper Baram. Their facial expressions were grave, and in their hands they held burning white candles. The monotonous throbbing of the big tree drum and the *seperut* posts cut from gleaming soft wood fibres affirmed the solemnity of the ritual.

My first journey to the Penan had taken me to this location in the midst of the virgin forest on the Selungo, land where the Penan have been living for centuries, refusing to yield it to the loggers. On that day in November 2004, I met with the Penan elders to plant a sago palm in memory of the missing Bruno Manser. Many of those present had known Manser in person and had held him in great esteem on account of his reverence for their culture as well as the assistance he had given them in standing up to the loggers. One of the headmen said to me: "When Bruno Manser disappeared, it was for us Penan as if the sun and moon had vanished from the sky and there was no light there anymore."

To welcome me, Headman Bilong Oyoi from the neighbouring village of Long Sait danced the eagle dance, which imitates the flight of the majestic bird of prey. The headman raised his arms and tilted his head devoutly towards the sky. Then he jumped in the air and uttered a whoop of joy. After a soft landing, he swayed to the sound of the *sape*, the plucked instrument played by the indigenous peoples of Borneo. Before the Penan converted to Christianity, they believed that the spirit of the eagle determined their lives, which were constrained by many taboos. The eagle dance is a gesture of deference to their history and tradition.

Together, Bilong and I placed a young sago palm in the trench that had been dug for it. It would be a memorial to the lost rainforest campaigner. *Uvut* is the name the Penan give to this plant, central to their culture. In the days when they were nomads, it was more than their source of staple food. Its dried leaf stalks were also used as darts for hunting. For the Penan, the robust *uvut* is a symbol of their resistance to deforestation.

A fine mist drifted off the trees into the sky as the morning sun slowly cleared the air. The time had come for the Penan to make their speeches. "Without our blockades, this forest would already have been logged a long time ago," said Kelesau Na'an, the host of the gathering, "and then we would have suffered like the Penan lower down the Baram, where there is hardly enough timber left for their own building needs and where most of the wild boar have disappeared."

One after the other, the headmen reported on the situation in their communities and the activities of the loggers in the region. Many of them complained that the Taib government had punished their resistance to deforestation by denying them identity cards and state funds for development. They had been left entirely to their own devices, but preferred that to losing the forest as their livelihood. The lack of healthcare and the state's lack of interest in providing schooling were acute problems for the Penan.

The next person to address the gathering was Melai Nak, headman of Ba Tik. He had walked several days to Long Kerong to call for help from the assembled headmen. The Shin Yang logging company was getting ever closer to his village with bulldozers and chainsaws. Melai was afraid that it would be only a matter of weeks until the loggers advanced to his village's drinking water catchment area. He was also fearful of what would happen to their hunting grounds if the loggers were not stopped.

The headmen debated how to proceed and decided to send a written warning to the manager of the timber camp, signed with the thumbprints of all the headmen present—the traditional Penan form of signature. If the logging company failed to heed the warning, there would be a second warning and then a blockade. In order to support the small village of Ba Tik, it was decided that the communities from the Upper Baram would send a delegation to assist Headman Melai in preparing a blockade.

On this occasion, the dispute with the loggers had a positive outcome. A few weeks later, Shin Yang, under pressure from the threatened block-ade, withdrew from the communal territory of Ba Tik. After the ferocious conflicts of the 1990s, the logging companies now steer clear of confron-tation. They much prefer to lure the indigenous people with promises, such as building a road or paying small sums in compensation. If the in-digenous people offer less resistance, the loggers march in with their heavy equipment, and soon nothing is left standing in the villagers' forest except a few worthless stumps.

MAPS OF THE RAINFOREST

Ernst Beyeler received us in the office behind his gallery at Bäumlein-gasse 9 in Basel. Here the famous art dealer and gallery owner had been buying and selling the works of Pablo Picasso, Wassily Kandinsky, and Alberto Giacometti for many decades. His legendary art collection in Riehen, Switzerland, forms the core of the Fondation Beyeler—the mu-seum endowed by him and his wife, Hildy. On that particular day, how-ever, the aging art dealer was looking at a blowpipe from Sarawak.

With puffed cheeks, he blew into the pipe made of rainforest hard-wood. A dart shot out, and, in a fraction of a second, was embedded in the wall, from which it took much effort to dislodge it. Beyeler laughed at his unexpected success with the Penan hunting weapon, and his blue eyes lit up.

Already at the beginning of his career, Beyeler, who died in 2010, had a great appreciation for indigenous art from Africa and Oceania. He was fascinated by the parallels with abstract contemporary art. What is less well-known is that the Basel gallery owner also made crucial contributions to maintaining the tropical rainforest through his foundation *Art for Tropical Forests*.

On that warm spring morning in May 2006, Joe Jengau Mela and Mutang Urud had travelled from Sarawak to Basel to present a blowpipe as a gift to Ernst Beyeler. "In our culture we only give this weapon to great personalities, whom we respect and honour," was written in an accompanying letter bearing the thumbprints of ten Penan headmen. "A poison dart from this weapon can kill any animal within minutes. With this gift, we express our gratitude and immense recognition for your support in our struggle for the rainforest and our land rights."

Without the support of Ernst Beyeler's foundation, the Bruno Manser Fund could hardly have survived the critical phase that followed its founder's disappearance. Between 2002 and 2007, the Beyeler foundation contributed more than half a million Swiss francs to finance the community mapping of the Penan territories in the Sarawak rainforest. This is the Bruno Manser Fund's most important project and involves much more than mapping the traditional Penan habitats in the rainforest. It extends to more detailed cultural documentation as well, including systematic transcription of the oral history of the mapped communities.

A team of Penan cartographers trained by the Bruno Manser Fund has been working in cooperation with the Penan villages to produce maps of the traditional land that they want to protect against further clearance. The first step involves entering all the known names of rivers, streams, mountains, and watersheds on sketches. Then the Penan indicate the locations where there are particularly good hunting grounds, occurrences of useful plants, as well as cultural sites, such as their ancestors' graves.

For the second step, the cartographers take to the forest with selected Penan elders to determine the borders of the community's territory and to record the GPS coordinates of all the important features of the land. It can take several weeks to gather the field data, during which time the mapping team remains in the jungle, far from the villages. Once the Penan have returned from the field, the data is entered into a computer and transmitted to Switzerland, where it is used to generate maps.

Since the beginning of the mapping project in 2002, the Penan have surveyed thousands of square kilometres of rainforest and produced maps that are amongst the most detailed of indigenous land anywhere in the world, recording the names of 5,200 rivers and streams, 1,800 other topographic features, as well as the locations of 513 poison-dart trees. At the same time, the histories of dozens of Penan communities have been recorded, transcribed and, in part, translated into English. Taken together, this wealth of cultural data is irrefutable evidence that the Penan have used the traditional forests for at least the period of their collective memory, proved by their detailed knowledge of their territory.

On the basis of these maps, six land rights claims had been filed before the end of 2013, covering 3,600 square kilometres of rainforest and farmland near the headwaters of the Baram, Tutoh and Limbang rivers. At the time of writing, the courts in Sarawak have not yet ruled on any of these cases.

KELESAU'S DISAPPEARANCE

In autumn 2007, disquieting news reached the Bruno Manser Fund from the Penan territory. Headman Kelesau Na'an of Long Kerong, one of the four plaintiffs and one of the most important witnesses in the Penan land rights claim from the Upper Baram, had disappeared. The nearly-80-year-old headman had last been seen near his village on 23 October 2007. He had told his wife that he wanted to check a number of traps that he had set in the forest to catch small wild animals. He never returned, and, despite sending out a search party immediately, the Penan were unable to find him.

Several weeks elapsed before they issued a public alarm. The incident reminded them of a case that had occurred in 1994, when the pastor of the neighbouring village, Ba Karameu, had disappeared after a dispute with logging company workers. The Penan had found him four weeks later, lying in a stream with his stomach slit open.

Kelesau's disappearance also happened at a time of heightened tension with the logging industry. Since April of that year, the Samling logging company had deployed security officers in the district to dismantle a blockade that Penan from the village of Long Benali had erected on a new logging road in the Upper Baram. The Penan, however, successfully withstood the logging company's attempts. In May, the Taib government had sent Samling on a mission to obtain Penan cooperation.[11] In June, Kelesau was threatened after logging company surveyors had been blocked by Penan protestors. The headmen of Long Kerong and Long Sait reacted with a written complaint to the manager of the nearby Samling logging camp.[12]

In the months that followed, Kelesau was repeatedly visited by a Samling envoy. She offered to provide assistance for his village and tried to persuade him to withdraw the land rights case. The headman, however, remained resolute and rejected the company's offers.

Two weeks after the headman's disappearance, a helicopter with two mysterious visitors landed in his village and offered the astounded inhabitants an outright gift of 30,000 ringgits (just under 10,000 US dollars). The two claimed they wanted nothing in exchange.[13]

Shortly before Christmas 2007, a Penan hunter found Kelesau's remains on the bank of a Selungo tributary, approximately two hours walk from Long Kerong. He recognised the corpse immediately from the missing headman's necklace and wristwatch. Kelesau's hand was broken, and his *parang* (machete) was gone. The Penan recovered the headman's remains and took them back to the village, where they were buried a few days later. On 3 January 2008 they reported the matter to the police in Miri and demanded an investigation. It was not, however, until after the Bruno Manser Fund made the incident public and international media took up the case that the Malaysian police, after waiting two months, had the dead headman exhumed.[14] They came to the conclusion that Kelesau had died a natural death.[15] Despite this, the Penan remain convinced that their tenacious headman was the victim of a violent crime.

SEXUAL VIOLENCE AGAINST PENAN WOMEN

An appalling discovery was made by the French journalist Andrea Haug in 2008 while filming a documentary about the Bruno Manser Fund's community mapping project. Accompanied by her brother and cameraman, Johann Haug, she was the first journalist to visit a remote village in the Middle Baram, where much of the forest had already been cleared. A 50-year-old Penan woman named Juma (name changed) broke her silence over what had been happening in her village, which was located not far from the logging camps of the Samling and Interhill companies.

"I, Juma, would like to inform the public that we Penan women are being sexually abused by loggers," she told the French journalist. She recounted shocking scenes that happened regularly in her village: A group of drunk loggers would arrive in the village unannounced in an off-road vehicle and immediately begin looking for women to assault. They would pick, in particular, on young women, including pregnant ones and girls who were scarcely thirteen.

"When we hear the vehicles coming, we drop everything and run away into the forest," Juma said. "We have complained to the managers of the logging camps and the police. We have never received any answers to our complaints, and the police tell us that we have just invented the stories. We are still waiting for a police officer actually to come to our village."

Andrea Haug had just returned from her journey to the Baram and had arranged to meet me in the coastal town of Miri. There she told me of her shocking interview with Juma. I judged the accusations against the loggers to be too serious to publish without verification. A Swiss doctor, who at the time was working as a volunteer for the Bruno Manser Fund, and was travelling in Sarawak, investigated the accusations. Once he had conducted interviews with the victims and confirmed their statements, the women's call for help was made public on 15 September 2008. Emails were addressed to the Malaysian government. Volunteers sent postcards to the Malaysian ambassador to the United Nations in Geneva.

There is hardly any other campaign organised by the Bruno Manser Fund in recent years that has had such a significant impact on the Malaysian public. Malaysia's largest English-language newspaper, the *Star*, took up the story, carried out its own research on the spot in Sarawak and published a lengthy report at the beginning of October 2008. It was written by Hilary Chiew, a courageous journalist from Kuala Lumpur.[16]

A Taib government spokesman challenged the statements made by the Penan women, and claimed they had all been fabricated. Next, an enraged Taib himself appeared before the media and accused the Bruno Manser Fund and other NGOs of "sabotage" against the state of Sarawak and against his development policy. He also demanded that the media who had reported on the affair should issue an apology and withdraw their stories.[17] Taib's deputy, Alfred Jabu, even attacked the Bruno Manser Fund (BMF) directly, saying the reports on sexual abuse of Penan women were "a great lie by BMF to the world".[18] The daily *Borneo Post* published a long article featuring a politician who accused Bruno Manser of having made a Penan woman pregnant during his stay in Sarawak, without offering any evidence.[19] With the exception of the Internet, all the media in Sarawak remain firmly in the hands of the Taib regime and the logging companies. Taib's youngest daughter, Hanifah, is director of the only private radio station, *Cats FM*, and of one of two English-language dailies, the *New Sarawak Tribune*.[20] The *Borneo Post* is owned by the KTS logging company.[21] It is virtually impossible to hold any public debate on politically sensitive issues.

The public uproar in Malaysia over the sexual abuses had mounted, however, and it was no longer possible simply to sweep the matter under the carpet. Malaysia's then Minister of Women, Families, and Community Development, Ng Yen Yen, raised the issue at a meeting of the federal cabinet in Kuala Lumpur, and commissioned an inquiry,[22] which was published a year later and confirmed the accusations made by the Penan women.[23] The report's findings were disseminated throughout the world and were taken up by international media, such as the *Washington Post*.[24] One year later, an independent fact-finding mission uncovered seven further cases of sexual violence against indigenous women and girls in the Baram region.[25]

Samling, which was mentioned by name in the *Star*, took action to sue the newspaper and the journalist, Hilary Chiew, for defamation. A leaked document shows, however, that behind the scenes the logging company did indeed take the matter seriously. In a memo with the title of "Rape case of Penan women", the responsible Samling manager instructed all employees not to enter Penan settlements or to give lifts to Penan without the company's authorisation.[26] Despite all this, up to the time of writing, not one criminal prosecution has been brought against the alleged perpetrators, who remain under the protection of the logging companies, the Sarawak state government, and the police.

07

OFFSHORE BUSINESS

International banks help Taib and his agents access capital markets. When it comes to the payment of kickbacks and to laundering money, timber barons seek discreet firms in the financial sector. The trail leads to Switzerland, Wall Street, and the City of London.

A MEMORIAL TREE IN FRONT OF CREDIT SUISSE

It was a normal Friday morning on 23 February 2007, and a cold north wind was blowing along Zurich's Bahnhofstrasse. Two dozen young people were marching slowly towards the Paradeplatz. On their shoulders they carried a pine trunk, on which images of rainforest animals and plants had been carved as a memorial to the missing Bruno Manser. The short procession was led by Janine Manser, the environmentalist's 20-year-old niece, who carried a sign saying: "Stop destroying the rainforests. Credit Suisse must stop financing the Samling timber group."

The February sun slanted across the roofs and warmed the neoclassical façade of the bank's headquarters on the Paradeplatz, but it remained a cold winter's morning as the protestors stood the heavy tree up at the entrance. Bruno Manser had disappeared in May 2000 on territory licensed to the Malaysian Samling Group, which remains one of the key players in the destruction of the Penan rainforests. Seven years later, Credit Suisse was leading a syndicate organising the stock market launch of the tropical timber corporation in Hong Kong—a slap in the face for the Manser family and the numerous Swiss friends of the missing human rights campaigner, who were demanding the bank's withdrawal from the controversial transaction. "We are dumbfounded that a Swiss bank can participate in such a transaction," said Bruno's siblings Erich Manser and Monika Niederberger-Manser, speaking on behalf of the whole family. "The Samling company is destroying everything to which our brother Bruno was committed for many years."

The bank had, however, made up its mind. "Credit Suisse is committed to Samling's initial public offering," were the words with which Tobias Guldimann, the bank's chief risk officer, ended a meeting with the Bruno Manser Fund and the Society for Threatened Peoples an hour later. With these words, the Zurich banker dispelled all doubts; the big Swiss bank placed its short-term financial interests above the preservation of tropical rainforests.

As "global coordinator", the investment bankers at Credit Suisse were setting out to raise US$ 280 million of new capital by the beginning of

March 2007 on behalf of the owners of Samling Global through its stock market launch. In doing so, they were supported by the British HSBC and the Australian Macquarie Securities. The timber barons from Sarawak were to receive a fresh injection of funds to enable them to expand their production capacities, to decimate even more of the rainforest, and to pay off their debts. Around ten million dollars would go directly to Credit Suisse, where the employees responsible for the transaction were looking forward to sizable bonuses.[1] The Penan in the Borneo rainforest were to remain empty-handed. The Malaysian tax authorities would also see little money from the deal. Samling Global was registered in Bermuda, where there is no corporate taxation.[2]

Following the stock market launch, the Samling share price shot up. Within four months, it had risen from the issue price of 2.08 Hong Kong dollars to 3.40 Hong Kong dollars. But the success was short-lived. Even before the outbreak of the financial crisis in 2008, the share price began a steep descent, and the smiles were wiped from the investors' faces.

After five years on the stock exchange, the company's owners, the Yaw family, decided in 2012 to delist it. As a precaution, the Yaws had never abandoned their majority holding in the company and had only placed a minority of the shares on the market. The minority shareholders were forced to sell up and were offered a cash settlement of 0.76 Hong Kong dollars per share, a loss of 63% compared with the issue price.[3] By the time the Samling Global shares ceased to be listed on 20 June 2012, Credit Suisse had washed its hands of any involvement with the shares of the tropical timber group.[4]

As early as 2010, ethical considerations had led the Norwegian government to divest itself of its shares in Samling, worth 1.2 million US dollars at the time, and to place the company on a blacklist. The shares had been acquired by the Norwegian state pension fund, one of the largest institutional investors in the world. Prompted by the campaign by the Bruno Manser Fund and its partners, the Norwegian government mandated the pension fund's Committee on Ethics to investigate Samling's business activities. At a press conference on 23 August 2010, the Norwegian minister of finance, Sigbjørn Johnsen, stated: "The Committee on

Ethics has assessed Samling Global and concluded that the company's forest operations in the rainforests of Sarawak and Guyana contribute to illegal logging and severe environmental damage. I have therefore chosen to follow the recommendation of the Committee and exclude the company from the investment portfolio of the State pension fund."[5]

The Samling case is a particularly graphic example of how international banks close their eyes to the social and environmental consequences of their transactions when fast money beckons, with devastating effects on the rainforest. Without the know-how and reputation of major global financial service providers such as Credit Suisse and HSBC, the timber barons in Sarawak would never have had access to the international capital markets. Conducting business with the timber barons can, however, be attractive and highly lucrative for western bankers, be they based on Wall Street, in the City of London, or on the Paradeplatz in Zurich. The British organisation Global Witness estimates that HSBC alone has earned over 130 million US dollars in dealing with Sarawak's seven largest timber companies since the end of the 1970s.[6]

Samling caused a stir in the financial world for another reason too. Its owners, the Yaw family, have been investing their huge profits from tropical timber around the world (see chapter 8) not only in the timber business but in other sectors too, most notably in real estate. Chee Siew Yaw, one of the sons of the company founder, lived in California from 1986 to 2006 and rapidly became an important player in the property sector there. One of the companies he set up, SunChase Holdings, was soon involved in the "largest real estate partnership ever formed with the U.S. Government", and spent more than two billion dollars buying properties from the US government. Mountain House, an urban development in northern California under construction since 2001 and numbering approximately 10,000 inhabitants, was a SunChase Holdings project. In 2008, the town had become an epicentre of the US subprime mortgage crisis with its global repercussions.[7]

The proprietors of the Samling property business are suspected of corruption. *Sarawak Report*, for instance, revealed that in the early 1990s, Chee Siew Yaw gave mansions in Seattle worth a total of 9.6 million dollars to the Taib family for free.[8] Another questionable arrangement is the Taib family's 15% holding of a property development project owned by the Yaws in Malaysia—the luxurious "Desa Parkcity" in western Kuala Lumpur worth several hundred million dollars.[9] Both parties keep a veil of silence over any possible link between these gifts and the issuance of logging concessions in Sarawak.

CHEQUES WORTH MILLIONS FOR UBS

Corruption in the tropical timber business is also an issue in the Swiss Bank UBS's dealing with Musa Aman, chief minister of the Malaysian state of Sabah. It all began on 11 April 2006 when the young Malaysian Michael Chia (otherwise known as Chia Tien Foh), an agent for Musa Aman, walked into the UBS branch in the financial district of Singapore. Chia had with him eleven cheques totalling 15 million US dollars in his pocket. He took the lift up to the 18th floor of the UBS building in Temasek Boulevard and asked to see his client adviser. The businessman and his banker disappeared through a door into the air-conditioned consultation room. Soon, the whole matter was settled. Chia's cheques bore a fresh, clean stamp: "UBS AG. Singapore Branch. Reception" along with the signature of the bank employee. Two days later, the 15 million were credited to Chia's bank accounts at the Hong Kong branch of the big Swiss bank.[10]

Michael Chia is the key figure in a money laundering affair that led to extensive investigations by the anti-corruption and judicial authorities in Hong Kong, Malaysia, and Switzerland.[11] There are substantial grounds for suspecting that in just two years—2006 and 2007—more than 90 million US dollars in kickbacks from the Malaysian trade in tropical timber flowed through accounts held at the UBS in Singapore, Hong Kong, and Zurich.[12] That is probably only the tip of the iceberg. The complex transactions involved not only UBS but also the British HSBC and the Singapore OCBC Bank. On the basis of a complaint by the Bruno Manser Fund, the Swiss Attorney General opened a criminal investigation against UBS and unknown parties on 29 August 2012 on grounds of money laundering and suspicion of breach of the bank's duty of care.[13]

Chief Minister Musa Aman was behind the cheques deposited by Chia. Just as in Sarawak, in neighbouring Sabah the head of government has the last word over timber concessions and export permits. Another parallel with Sarawak is that in Sabah too it was primarily the chief minister who profited from clearance of the rainforest. Musa is the brother of Malaysia's foreign minister, Anifah Aman. Musa's wife is related to the family of Malaysia's federal attorney general, Abdul Ghani Patail. In Malaysian political circles that is as good as a guarantee of immunity from criminal proceedings.

IMPECCABLE SERVICE FOR THE TIMBER MAFIA

One principal witness of the Swiss UBS investigation states that "Musa Aman took a careful look at his colleague, Taib, in Sarawak, to see how he had managed to profit from deforestation. The most important difference is that Musa has only been in business since 2003, whereas Taib has been doing it for more than three decades. Taib has used this time to perfect his system of corruption. In Sarawak he also has control over much larger and more valuable forests, because most of the timber in Sabah was already felled many decades ago."

In a thin cotton shirt and without a jacket, the witness defied the cold November mist in the historic city centre of Berne. The man in his mid-thirties, whom we'll call Ling, had just been questioned by the Office of the Swiss Attorney General. He had offered to assist the Swiss investigators in sorting out the role played by the UBS in the money laundering affair. "No, no, I'll get by, I've experienced worse than this," he said in turning down my suggestion that he might buy some warmer clothes for his short visit to Switzerland. Anyway, dressed like that, the Rolex on his wrist stood out all the more.

Once we had arrived at the Kursaal in Berne, Ling produced a sketch to explain the routes taken by both the tropical timber and the profits associated with it. Monika Roth, a professor of law and the lawyer representing the Bruno Manser Fund in the UBS case, the journalist Clare Rewcastle, and I listened attentively. The date was 20 November 2012.

Ling himself was active in the timber trade and was familiar with every detail. "I can give you a precise explanation of the system of corruption and money laundering invented by politicians, logging companies and bankers," he told us. "The timber business functions according to the same principles everywhere in Southeast Asia. The very first opportunity for corruption is in the issue of timber concessions. Despots like Musa and Taib do not lose any opportunity to demand bribes, and issuing concessions is one of the best."

In this, the politicians pocket millions just for one signature—a bribe decides whether a logging entrepreneur will make it or not. Without state concessions, no one can harvest or process timber in Malaysia. No one who does not pay bribes has any chance whatsoever of getting into the timber business. That says a lot about an industry that lives from close ties with politics—and which, along with the corrupt governments, bears the main responsibility for the destruction of the tropical forests of Southeast Asia. In exchange for the payment of hefty bribes, the government and police simply avert their eyes if the logging companies cut down excessive volumes of timber, or fell protected species, or violate laws in any other way.

"The next possibility for corruption comes with the transport of the timber from the forests in the interior of the country to the export ports on the coast. Transport by truck is expensive and dangerous," Ling said. "For larger quantities, the timber companies use barges, requiring a government licence. Once again, politicians' palms need to be greased."

The UBS's client, Michael Chia, Musa's middleman, was also familiar with this business. His father had secured a monopoly from Musa to operate timber barges on various rivers in Sabah back in 1997—a veritable gold mine.[14] Musa Aman also takes his cut from permits for the next transport stage—export—in a similar way to Taib's brother Onn, who hands out export permits for cash (see chapter 5).[15]

THE UBS MUSA AMAN CASE

TIMBER GROUP
Plans to log
in Sabah

MUSA AMAN
Chief Minister
of Sabah
(Malaysia)

MICHAEL CHIA
Musa Aman's
nominee

1   Timber group pays bribe to Hong Kong account

2   Transfer of funds through various accounts

3   Wire transfers

4   Chia notified that a payment has arrived

5   Musa notified that a payment has arrived

6   Logging permit granted

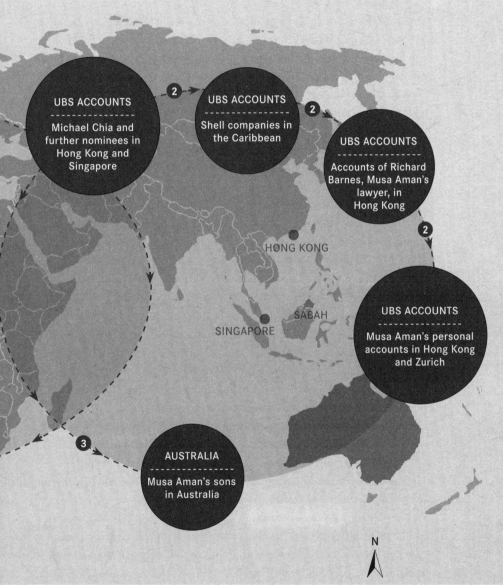

UBS ACCOUNTS

Michael Chia and further nominees in Hong Kong and Singapore

2

UBS ACCOUNTS

Shell companies in the Caribbean

2

UBS ACCOUNTS

Accounts of Richard Barnes, Musa Aman's lawyer, in Hong Kong

2

HONG KONG

SABAH

SINGAPORE

UBS ACCOUNTS

Musa Aman's personal accounts in Hong Kong and Zurich

3

AUSTRALIA

Musa Aman's sons in Australia

N

Source: BMF 2013

"When our logs are exported to be processed in countries like Japan, China, Korea and Taiwan, lots of money flows out of the country and never comes back again," Ling continued. "Our timber companies present forged delivery dockets with exaggeratedly low sums on them, so that it appears that they never make any profit in Malaysia. They arrange to have the invoices with the true sums paid to front companies in Hong Kong and Singapore. From there, the kickbacks are transferred directly to the foreign bank accounts of the Malaysian politicians. In that way, the kickbacks never touch Malaysia itself."

The money may well be invisible in Malaysia, but it does appear in bank accounts in other countries. Estimates by the World Bank put the annual non-reported profits of the global timber business at between 10 and 15 billion US dollars—money that is systematically diverted from the national economies of the timber's countries of origin.[16]

"UBS is only one of many banks that have participated in this business," reported Ling. "All the international banks with activities in Southeast Asia know about this business, and many have profited from it." Then he added: "I don't like saying this, but UBS provided us with an outstanding service."

Ling explained how Michael Chia's client adviser and the latter's boss at UBS in Singapore must have known from the very beginning that Chia's money was anything but clean. Previously, Chia had transacted business through the HSBC branch in Singapore until things became too hot for HSBC, and it called on Chia to close his accounts. UBS welcomed the Chief Minister Musa's envoy with open arms.

The bribes from the logging companies were paid in instalments of a few hundred thousand dollars, some of them with the endorsement "Deposit for logging concession", but on occasions up to two million dollars were transferred at one go. Chia made sure that the balances on his various accounts were never too high to be conspicuous. He moved the money through a web of companies and accomplices—all with accounts at UBS in Hong Kong. UBS also organised for Chia the companies needed for the cover-up, each with domicile in the British Virgin Islands, with names like CTF International, Blisstop, and Zenique Investment.

The final destination for the transactions totalling more than 90 million US dollars was either Musa Aman's personal lawyer, Richard Barnes, or Musa's own personal accounts with UBS in Hong Kong and Zurich.[17] UBS must have realised that Musa and his lawyer were "politically exposed persons" (PEPs), and should have refused to accept his dubious monies. The bank claims that it reported Musa's dodgy businesses to several countries' anti-money-laundering authorities in 2008 but it is obvious that for the two previous years, UBS had done nothing about it.[18] That was to come back to haunt the big Swiss bank during the Swiss criminal investigation later on, especially since at that time it was also involved in the global manipulation of the Libor rate and illegal transactions in both the USA and France.

By the time things had got that far, however, Musa's money had long since left UBS and, if insiders are to be believed, has now been invested with a Dutch bank instead.

DEUTSCHE BANK'S MURKY SECRET

In 1992, shortly after his return from Malaysia in 1992, Bruno Manser opened an account at Deutsche Bank in the town of Lörrach, just on the German side of the border with Switzerland at Riehen. The reason for the account was to make it easier for benefactors in Germany to make their personal contribution to saving the Sarawak rainforest. The Bruno Manser Fund kept that donation account active for twelve years, and relations with the bank's branch office in Lörrach were always friendly. The account number was published with every appeal for funding by the Bruno Manser Fund and on the Internet.

When I found a letter from Deutsche Bank in the Bruno Manser Fund's letterbox in the middle of October 2004, I initially thought it was just routine bank correspondence. When I opened it, I was surprised to read: "According to clause 19 paragraph 1 of our General Terms and Conditions, we have the right to discontinue our business with you at any time. We hereby inform you that we are making use of this right and your account will be closed on 19 November 2004. Please make the necessary arrangements, considering that we will not be conducting any banking transactions for you after that date."[19]

Thinking someone had made a mistake, I telephoned Deutsche Bank. An employee at the Lörrach branch was non-committal. He claimed the European Union had adopted new provisions making it impossible for them to run bank accounts for associations based abroad. When I asked for the references of those new provisions, no answer was forthcoming, instead I got curt advice to "phone our Private Banking unit in Duisburg—they're the responsible ones." I doubted that Private Banking could be responsible for an association's account with a balance of a few hundred euros at a bank that managed assets running into billions.

The gentleman in Private Banking in Duisburg listened politely to my request, typed the account number into his computer and commented briefly: "Unfortunately, I am not able to give you any information regarding this account closure."

"Well, is there any reason for the closure?" I asked.

"Yes. There is a reason, but no one is going to tell you what it is."

The contact with Duisburg at least resulted in an extension of the deadline for closing the account. That gave the Bruno Manser Fund sufficient time to open a new account in Germany. It did so at the Deutsche Postbank in Nuremberg. That posed no problem, and no one there had heard anything about the purported new European Union provisions.

I decided to send a written protest to Deutsche Bank's head office in Frankfurt. I also sent a copy to Josef Ackermann, the Swiss chairman of the bank. Like most Swiss people, Ackermann would be familiar with the name Bruno Manser—his 60-day hunger strike in front of the federal parliament building in Berne had made Manser a household name.

Shortly before Christmas, I wrote to Germany's largest bank: "We are aware that Deutsche Bank is amongst the largest foreign investors in Malaysia and is at present working intensively on further expanding its market position in Southeast Asia. We cannot therefore exclude the possibility that political considerations may have played a part in your decision to terminate our account."

The cordial reply was addressed to Bruno Manser himself, who had already been missing for four years. It promised to look into the matter and thanked me for my letter "also in the name of Dr. Ackermann". Two weeks later, on 6 January 2005, Deutsche Bank sent its final decision: "We are not required to give any reason regarding our decision to use our right to terminate an account. We would like to thank you for your cooperation over the past years. [...] With warmest New Year greetings."[20]

OFFSHORE DEALS ON THE TROPICAL ISLAND

At the time of this correspondence with Joe Ackermann's subordinates, I did not yet know that Deutsche Bank had just granted a loan worth 135 million US dollars to the Taib government.[21] Six months after that, Deutsche Bank issued a highly profitable government bond of over 600 million US dollars for Taib in the Malaysian offshore financial centre of Labuan.[22] Organising loans for the Taib government meant significant short-term profits for the investment bankers from Frankfurt. It is understandable that under these profitable conditions no one was going to want to risk spoiling Taib's mood by maintaining an insignificant donation account for an association set up by his adversary.

Labuan is a tropical island 8 kilometres off the north coast of Borneo and can be reached quickly by express ferry. The former British Crown colony used to be a sleepy island state, which never really managed to attract any visitors apart from the occasional diving tourist. Then, in 1990, Malaysia's prime minister, Mahathir Mohamad, had the idea of turning Labuan into an offshore financial centre with only a minimum amount of regulation for the financial industry. Since then, nearly all large banks have opened branches in Labuan—not just Deutsche Bank but also UBS, Credit Suisse, HSBC, Goldman Sachs, Citibank, and many more. For Taib, it meant a financial centre on his front doorstep, where he was able to do business with the big players of international finance with virtually no supervision.

The names of many of these banks keep cropping up in connection with Taib's worldwide financial transactions. Apart from Deutsche Bank, UBS and Goldman Sachs have also had their fingers in Sarawak government bonds. In December 2004, UBS floated a loan worth 350 million US dollars for the Taib regime in Labuan and Luxemburg.[23] In 2011 and 2012, it was Goldman Sachs' turn, and it issued two loans of 800 million US dollars each for the Taib government.[24] According to the *Wall Street Journal*, Goldman Sachs reportedly earned more than 50 million US dollars with the first of these two loans alone. Malaysian government bonds are regarded by the US bank as one of its most lucrative lines of business.[25]

The Taib government used part of the 2004 UBS loan (the so-called "Sarawak Corporate Sukuk") to acquire a struggling manufacturing plant from semi-conductor producer 1st Silicon in Kuching.[26] Two years later, the government sold 1st Silicon to the German X-Fab group, at an estimated loss to the Sarawak state of 2.5 billion ringgits (roughly US$ 800 million). Had Taib deliberately overpaid 1st Silicon and pocketed the difference for himself? When an opposition member of the state assembly (Sarawak regional parliament), Chong Chieng Jen, asked critical questions about the 1st Silicon deal in 2010, the assembly's speaker instantly suspended him.[27]

Chong, a youthful-looking lawyer from Kuching, has been critical of the Taib government's financial dealings for many years. Since 2004, the energetic politician has represented the Kuching constituency in the Kuala Lumpur parliament. In 2006, he was elected to Sarawak's regional parliament for the Democratic Action Party (DAP). Chong is also vice-president of DAP Malaysia, which has a leftist platform. The party with a red rocket as its symbol has committed itself to fighting corruption and it follows that Taib's grab for state revenues has been a thorn in the flesh of the DAP for a long time.

In early 2013, the DAP discovered that over a period of eight years the Taib government had arranged for over 11 billion ringgits (3.4 billion US dollars) to disappear into a secret fund, behind which the opposition suspects companies and front organisations belonging to the Taib clique. In addressing the Sarawak state assembly, Taib, in his capacity as minister of finances, went no further than general statements to the effect that this fund was there for Sarawak's "development".

"Secrecy about the way resources are spent contradicts one of the fundamental principles of parliamentary democracy," comments Chong, as a member of the opposition.[28] Back in 2009, the DAP was already criticising the fact that three of the ministries headed by Taib control half the government's spending, while the remainder is divided over the other ten ministers.[29] Without transparency, there is no way of knowing whether a significant part of this money ends up in the pockets of Taib, his family, and his political allies, who invest their illegally-acquired assets not only in Malaysia but in the financial and real estate sector all around the world.

That huge sums are diverted out of Malaysia's national economy is confirmed by Global Financial Integrity, a non-profit organisation with its headquarters in Washington D.C., which campaigns for transparent capital markets and barriers to stop the flight of capital from developing and newly-industrialised countries; it estimates that between 2001 and 2010, approximately 285 billion US dollars of unreported earnings flowed out of Malaysia into the international capital markets. That made Malaysia the country with the third-largest capital flight anywhere in the world, after China and Mexico, whose national economies are several times larger.[30] Only part of this money originated in Sarawak, but even that amount is large enough to be extremely attractive to international banks.

PECUNIA NON OLET

Deutsche Bank entered into particularly close ties with the Taib family through the Malaysian financial business of K & N Kenanga Holdings. Kenanga is the Malaysian name for *Cananga odorata* (or "ylang ylang"), a tropical blossom with a fragrance similar to jasmine that is used in famous perfumes such as Coco Chanel's Chanel N° 5. The Roman saying *Pecunia non olet* ("money does not stink") is particularly relevant for the

Kenanga Group, for they are not interested in perfume: They are concerned only with money and profit.

K & N Kenanga Holdings has equity totalling 612 million ringgits (just under US$ 190 million), and its biggest shareholder, with 25.1%, is the Taib-owned CMS Capital. Deutsche Bank has a holding of 13.8% through its subsidiary, Deutsche Asia Pacific Holdings in Singapore. The members of its board include, amongst others, Taib's son-in-law, Syed Ahmad Alwee Alsree; Richard Curtis, CEO of CMS; plus three current and former Deutsche Bank managers.[31]

One of the group's subsidiaries, Kenanga Deutsche Futures—of which Deutsche Bank holds a 37% share—is active as a broker in equity trading on the Malaysian exchange.[32] Classical banking is another of the activities of the Kenanga group. At the end of 2011, Kenanga administered more than 2.5 billion ringgits (roughly 760 million US dollars) in client assets.[33]

In the recent past, Kenanga has expanded its investment banking activities and created subsidiaries in Sri Lanka, Vietnam, and Saudi Arabia. Kenanga also raised funds for Taib's favourite project in Sarawak, the construction of twelve new dams in the framework of SCORE, the "Sarawak Corridor of Renewable Energy" (see chapter 9). In 2011, Kenanga raised 2.5 billion ringgits (760 million US dollars) for Sarawak Energy, the Taib government's state electricity concern and owner of the controversial dam projects.[34] Deutsche Bank avoids public view in matters of such highly sensitive projects; inquiries from the media and a letter from the German Society for Threatened Peoples were left unanswered by the Frankfurt bank.

Deutsche Bank's close ties with Kenanga show that Germany's leading bank does not shy away from business with the Taib family. But the important question is whether the Kenanga Group is involved in laundering money acquired illicitly by the Taibs. In 2011, the supervisory authority of the Kuala Lumpur exchange imposed a fine of 200,000 ringgits (60,000 US dollars) on Kenanga Deutsche Futures for infringements of the Malaysian money laundering legislation.[35]

There is another business connection between Deutsche Bank and the Taib family that has not yet been explained, owing to the lack of cooperation from Frankfurt. According to whistle-blower Ross Boyert, various Taib properties in the USA are held through a company called Sogo Holdings, which is domiciled in Jersey, a British Crown dependency in the Channel Islands.[36] Sogo Holdings was also involved in a 20-million-dollar loan to Taib's Sakto property group in Canada (see chapter 1).[37] Sogo Holdings' shares in Jersey are all issued to offshore trusts of Deutsche Bank. One third is held by the Deutsche Bank subsidiary on the neighbouring island of Guernsey and two thirds by the Deutsche Bank International Trust Co. on the Cayman Islands in the Caribbean.[38]

Following questions addressed by the Bruno Manser Fund to the German government in 2011, BaFin (the German Federal Financial Supervisory Authority) took a close look at Deutsche Bank's relations with the Taib family.[39] Its examination dealt with "compliance with the due diligence duties laid down in the anti-money-laundering law" and also "the internal security measures set up by the financial institute". BaFin concluded that there were "no grounds" for action by the regulatory authorities.[40]

TASTY LITTLE FRANKFURTERS

In order to find out more about Deutsche Bank's relationship with the Taib family, the Bruno Manser Fund purchased a number of shares in the bank in the spring of 2013. Like all shareholders, it has the right to attend the Annual General Meeting and to put questions directly to the top management. The big day came on the morning of 23 May 2013. Mutang Urud from Sarawak and I put our names down on the list of speakers in the huge auditorium called the "Festhalle" at the Frankfurt exhibition grounds. Our aim was to demand that the Deutsche Bank management account for itself as far as the Taib transactions were concerned. Of course, I also wanted to know why the Bruno Manser Fund's account had been closed and whether, nine years later, we would be able to open an account at Deutsche Bank again.

Lunch comprised tasty frankfurter sausages with mustard and potato salad. Blue paperweights in the shape of the Deutsche Bank logo held the white tablecloths in place. Hundreds of grey-haired minor shareholders milled around the buffet.

After six hours of waiting, by which time large numbers of participants had already left, Mutang and I were finally allowed up to the podium. We had four minutes in which to present our case before the chairman switched the microphone off. A few minutes later, the co-CEO, Jürgen Fitschen, reacted to our questions. His reply was polite, more or less non-committal, and did not go beyond commonplaces. Actually, I had not been expecting more than that, but what is more important is the legal consequence of our appearance. No one can continue to claim that Deutsche Bank is conducting business with the Taib family "in good faith" and is unaware of our accusations of corruption.

The Bruno Manser Fund was treated to a much clearer message from Deutsche Bank by email a few days later. It gave an unambiguous answer to the question as to whether or not the BMF was a welcome client again:

Re: Resumption of business relations
Date: 27 May 2013 13:35:26 CEST
To: info@bmf.ch
Classification: Confidential
Dear Dr. Straumann,
Thank you for your inquiry concerning the resumption of a mutual business relationship.
I have to inform you, however, that we at Deutsche Bank are not interested in a resumption of business dealings with the Bruno Manser Fund. I regret that I cannot give you a more positive decision, but remain with kind regards,
(...)
Deutsche Bank Privat- and Geschäftskunden AG.[41]

SWISS CONNECTION?

Are Swiss banks also managing assets running into millions for Taib and his clique? That is a question that had been going through my mind for several years ever since a source from Malaysia had claimed that Taib used to fly to Geneva in his private jet whenever he needed dental treatment. After Switzerland had frozen millions in assets held in the names of the ruling families of Egypt, Tunisia, and Libya in the course of the Arab Spring early in 2011, the Bruno Manser Fund informed the then federal councillor and Swiss foreign minister, Micheline Calmy-Rey, of rumours circulating in Malaysia, according to which UBS was managing assets on behalf of the Taib family. In addition, Taib's niece, Elia Geneid Abas (who received 10,000 hectares of state land from her uncle), had been married to a Swiss since 2010 and maintained close relations with Switzerland. The Bruno Manser Fund therefore called on the Swiss Federal Council in March 2011 to freeze any assets that Taib might have in Switzerland.[42]

That would, unfortunately, not work, was the answer from the foreign minister. It was true that the Federal Council had powers to freeze the assets of foreign potentates, but it would only make sense to use those powers if there were real prospects of returning such assets to the countries concerned under a request for judicial assistance. "As long as the persons concerned are still in power, however, it is unlikely that the necessary requests for judicial assistance will be forthcoming," and hence it would not be possible for the courts to probe the possible criminal origin of such assets. The federal councillor did, however, forward the BMF's accusations to FINMA (the Swiss Financial Market Supervisory Authority).[43]

The news that the Swiss president had issued a statement on the Taib assets was seized by the online media in Malaysia and led to official reactions. In the newly-elected state assembly, Taib announced that he, in turn, would also make a statement. All the journalists were eagerly waiting when Taib announced to virtually no one's surprise that he was not corrupt and that he had no money in Switzerland.[44] As a direct consequence of the BMF campaign, the Malaysian anti-corruption commission (MACC) opened an investigation into Taib on 9 June 2011.[45] Three years later, at the time of writing, the commission has not yet released any findings from its investigations. By contrast, new accusations against Swiss banks and against Taib have come from within the Taib family itself.

"My former father-in-law, Taib, is the richest man in Malaysia and one of the richest men in the whole of Southeast Asia." The elegant woman with the beige headscarf and pink lipstick took an oath before the Sharia court in Kuala Lumpur, swearing to tell only the truth. Then the 49-year-old Shahnaz Abdul Majid, sister of a famous Malaysian pop singer, embarked on her disclosure of everything she knew about the family of her ex-husband, Mahmud Abu Bekir Taib.

"My ex-husband has 111 personal bank accounts with balances totalling more than 700 million ringgits [210 million US dollars] in Canada, the USA, the Caribbean, France, Monaco, Switzerland, Luxemburg, Hong Kong, and Malaysia. He also owns properties in Sarawak, Peninsular Malaysia, and scattered all around the world. He is a director of 112 firms and a shareholder in fifty businesses. Whenever he travels abroad, he only ever uses a private jet."[46]

Shahnaz Abdul Majid had been married for 19 years to Taib's elder son, Abu Bekir, before he abandoned her and ran off with a young Russian blonde. At first, she had hoped he would come back to her, but, having waited until December 2012, all that Shahnaz now wanted was her share of the Taib assets—and to have nothing more to do with Abu Bekir. The former director of the Taib CMS business decided to confront the politician's family in public.[47] What she had to tell the court in the divorce hearing had the Taibs trembling—and the Malaysian public dumbstruck.

Shahnaz began by demanding a *mutaah* ("severance gift") of 400 million ringgits (roughly 120 million US dollars) from her ex-husband as a settlement for her divorce and for her pain and suffering. Later on, she reduced her demand to a modest 100 million ringgits (30 million US dollars)—a minor sum for Taib's son, for whom, as she put it, "these hundred million are worth precisely what ten cents are worth to other people." In putting her case, Shahnaz was counselled by a shrewd Indian lawyer, Dr. Rafie Mohd Shafie, who, in examining his client before the court, obviously revelled in extracting the details of the Taib family assets.[48]

Shahnaz's testimony also sent shock waves to Switzerland. However, her accusations could not be substantiated. An examination carried out by the office of the Swiss Attorney General in 2013 showed that only one of the Swiss banks mentioned by Taib's daughter-in-law had had a business relationship with the Taib family. Their bank accounts had already been closed in 1999.[49] Contrary to the Musa Aman case, the Swiss prosecutors did not open a criminal investigation and ordered not to proceed with the matter.

08

TRAIL OF DESTRUCTION

The timber barons of Sarawak have thrived under the Taib regime. Today, they are among the most important players in the global tropical timber business. Multinational companies, like Rimbunan Hijau and Samling, are involved in legal and illegal deforestation all around the world. At a breathtaking pace, they are decimating virgin forests in Southeast Asia, Africa, South America, and Australia.

A RAINFOREST TRIBAL CHIEF ON ZURICH'S BAHNHOFSTRASSE

The necklace made of out of the rinds of rainforest fruits hanging over the loosely-knotted tie identified the gentleman in the dark suit as a *Toshao*, the tribal chief of an indigenous village in the Amazonian state of Guyana. David Wilson is the head of the village of Akawini on the river of the same name, where some 800 people live, about 70 kilometres to the north of the nation's capital, Georgetown. The young dark-eyed chief had travelled 7,700 kilometres from his home to the Bahnhofstrasse in Zurich, where he was confidently addressing journalists in defence of his village against the Malaysian timber group Samling and its financial backer, Credit Suisse.

"Samling cleverly deceived us," the *Toshao* reported. "We allowed a retired school teacher into our forest to cut down a few trees. But it was all a planned fraud. His small timber company, Interior Wood Products, was just a front for Samling."

Two months after Samling was listed on the Hong Kong stock exchange, rainforest dwellers from Sarawak in Southeast Asia and Guyana in South America met in Zurich in May 2007. The representatives of the indigenous peoples had been invited by the Bruno Manser Fund and the Society for Threatened Peoples to inform the Swiss public about the machinations of the Malaysian tropical timber group Samling. Despite international protests, Credit Suisse had successfully launched Samling on the stock market. Now, the indigenous peoples were demanding ten million US dollars from the big Swiss bank as compensation for the destruction of their habitat. It was a purely symbolic demand, given that the rainforest dwellers had no leverage over the loggers and their Swiss bankers.[1]

"One day, the director of Interior Wood Products came into our village, accompanied by a number of people from the government," *Toshao* David Wilson reported. "He brought a contract with him and told us that he had the government's approval to fell trees on our territory. They then divided us up into various groups and gave us five minutes to read the contract, although we didn't understand the legalese." The inhabitants of Akawini were then compelled to sign the agreement with the timber company. A few days after signing, workers from the Barama Company, a Samling subsidiary, turned up with large bulldozers, lorries, and excavators and began to cut down the forest.

"The Barama workers have no respect for the village inhabitants and the village council. They come into our village without our permission and hunt our game. We've told them to stop, but they keep on doing it. We urged them not to build any bridges over the Akawini River, which is our most important waterway and our source of drinking water. But they didn't abide by any of that," David Wilson continued. "The result of Samling's activity in our village was a typhus epidemic. Nothing like that had ever affected us in Akawini before."[2]

On his visit to Switzerland, *Toshao* David Wilson was also joined by the president of the Amerindian Peoples' Association, the lawyer David James, and Janette Bulkan, an anthropologist from Guyana.

Guyana, on the Atlantic coast of South America, covers a land area of 215,000 square kilometres, but it has only 770,000 inhabitants, who earn their livelihood mainly from mining and farming. The principal export products are gold, bauxite, sugar, and rice.[3] More than 85% of Guyana is covered in tropical rainforest, where over a thousand tree species, 1,600 bird species and 1,100 other vertebrate species are indigenous. The Guyana upland, the Guiana Shield—a topological feature that extends into the neighbouring countries of Venezuela, Brazil, Surinam, and French Guiana—is, with Borneo, one of the regions on Earth with the highest biodiversity. Jaguars, parrots and the curious hoatzin—Guyana's national bird—are just some of the truly fascinating creatures that live in this remote rainforest region.

Guyana, a former British colony, was granted independence in 1966 and is the only English-speaking country in South America. Some 30% of the Guyanese are Creoles (African Guyanese), descendants of the African slaves taken across the Atlantic by Dutch settlers to work on the sugar cane plantations. After slavery was abolished in 1833, the British colonial masters brought Indian workers to the plantations. Their descendants account for 43% of the population, whereas the proportion of indigenous peoples amounted to only 9% at the beginning of the 21st century.[4]

The Guyanese anthropologist Janette Bulkan, who is currently professor of indigenous forestry at the University of British Columbia in Vancouver, was also hard-hitting in what she had to say about Samling. "Samling has bigger logging concessions in Guyana than in Malaysia, more than 1.6 million hectares. In addition to that, Samling is operating illegally in 400,000 hectares of forest, for which the concessions were actually granted to other companies."[5]

Janette Bulkan, who was born in a sawmill, knows what she is talking about when she denounces the overexploitation of the rainforest. On account of her forthright criticism of political corruption in Guyana and the scheming of the logging industry, the scientist has earned many enemies in her home country, where she used to work in the Iwokrama International Centre for Rainforest Conservation and Development with the Makushi and other indigenous peoples.

The arrival of Samling and other Asian logging companies in the early 1990s upended the forestry sector in Guyana. "Asian businesses benefit from the inadequate state regulation of forestry," Bulkan says. "They have taken over under-capitalised concessions that were only moderately exploited, and things have now gone so far that the Asian loggers control 79% of Guyana's medium- and large-scale logging concessions."[6]

What particularly annoys Bulkan, she announced at our meeting in 2007, is that the Malaysians have not paid a single dollar in business taxes in Guyana since their arrival in 1991, despite the fact that, on the basis of a foreign direct investment contract with the Guyanese government, Samling benefits from tax-free fuel and other privileges. Through accounting tricks, Samling has never reported a profit in Guyana. The company claiming to be permanently in the red quickly became the biggest timber exporter, sending most of the wood to China and India in the form of unprocessed logs.[7]

An unequalled act of impudence occurred in 2006, when Samling's subsidiary, Barama, despite its involvement in illegal logging in a significant part of its operations in Guyana, managed, with the backing of WWF Guianas, to get the Forest Stewardship Council (FSC) label to certify its forestry management as "sustainable". In March 2006, the WWF boasted of a new record in forest certification, with the newly FSC-certified Samling operations on 570,000 hectares of tropical forest in Guyana constituting the largest area of FSC-certified tropical forest anywhere in the world.[8] The sustainability certificate was issued by SGS Qualifor, a South African subsidiary of the Swiss certification company SGS, which has its headquarters in Geneva.

It took painstaking research and determined questioning by Janette Bulkan and her husband, John Palmer, to demonstrate that Samling was not abiding by the strict FSC requirements and should never have been certified. When the FSC's associated accreditation body (Accreditation Services International) reacted to public pressure and looked into Samling's activities in Guyana, it backtracked totally and, in January 2007, cancelled Barama's FSC label.[9]

Despite the scandal surrounding the FSC certification of Samling in Guyana, Credit Suisse, along with HSBC and the Australian bank Macquarie Securities, still went ahead with the launch of Samling Global on the Hong Kong stock exchange two months later. The turning point only came with the international NGO campaign against Samling. At the end of May 2007, four weeks after the delegation from Guyana had appeared in Switzerland, public pressure against Samling had risen to such an extent that the timber corporation was obliged to withdraw its bulldozers from the forest around the villages of Akawini and St. Monica. It had been worthwhile for the indigenous people to protest against the destruction of their forest, and for *Toshao* David Wilson to make the journey to Zurich.[10]

In the autumn of that same year there were, at long last, political consequences and criminal sanctions against Samling's illegal operations in Guyana. Guyana's President, Bharrat Jagdeo, expressed strong condemnation of Barama and revealed that employees of the national forest service had provided a cover-up for the Malaysians' breaches of the law.[11] The matter ended in fines totalling more than two million US dollars imposed on Barama and various Asian companies for fraud, illegal log-ging outside of their concessions, and the forgery of timber export docu-ments.[12] Barama alone paid almost one million dollars, by far the largest penalty ever imposed in Guyana—and paid with little protest.[13]

At the time of writing, the Samling group has not won back the FSC certification for its subsidiary Barama in Guyana. In 2009, a disabused WWF broke off its dealings with the Malaysian company, which will not commit to sustainable forestry management at any price.[14]

SARAWAK, THE EPICENTRE OF DEFORESTATION

Samling's actions in Guyana are by no means an exception, but are the distressing norm for Sarawak timber barons. They have already been forced to withdraw their bulldozers from large areas in Cambodia, Papua New Guinea, and Brazil because of their illegal activities. All around the globe, a trail of destruction cuts through the red soils of the tropical rainforest. Anyone following this trail will in many cases find connections with Taib's logging state. The "Fair Land Sarawak", once a private colony of the Brookes, idealistic and paternalistic at the same time, has become the epicentre of tropical rainforest destruction under Taib's rule.[15]

Global Witness, an organisation based in the United Kingdom, esti-mates that four logging and plantation companies from Sarawak (Sam-ling, Shin Yang, WTK, and Ta Ann Holdings—all of them clients of the HSBC bank) are involved in clearing 18 million hectares of forests around the world or are transforming them into plantations.[16] To that must be added 8–10 million hectares of logging and plantation concessions in the hands of the firm Rimbunan Hijau, the biggest player in the global ex-ploitation of tropical rainforest. This company from Sarawak with sales in excess of a billion US dollars and a presence in seventeen countries on six continents diversified long ago from being a pure logging operation into a conglomerate, and is today involved also in the oil and gas busi-ness, tourism, the media, and information technology.[17]

Sarawak's logging and plantation industry leads the world in forest destruction—with devastating impact on biodiversity and the long-term survival of the indigenous peoples. What is happening to the rainforests of equatorial Africa? Rimbunan Hijau is busy chopping them down.[18] And the age-old trees in the paradise of Papua New Guinea's forests? No fewer than three timber companies from Sarawak are busy processing them as logs.[19] The ancient Tasmanian eucalyptus? Ta Ann is cutting them up into parquet for the Japanese market. Even in the dense coniferous forests of the Russian Taiga, the chainsaws of the Malaysian timber barons are roaring day in, day out. The former colonial subjects from Borneo have now become the colonialists themselves and surpass, in their ruthlessness, the example set by Europeans in the past.

What all these companies have in common is that they have their roots in the despotic state of Sarawak and that, under Taib's protection there, they developed a business model based on corruption, environmental destruction, and violation of the rights of indigenous peoples. They then put it into practice, and subsequently exported it to other parts of the globe.

A drive for profits combined with callousness as far as social justice and nature protection are concerned are the essential ingredients of this approach to business, which has brought great wealth to a small clique of individuals, but has brought deprivation of the rights and marginalisation of the many indigenous inhabitants who are the victims of this deforestation and suffer permanent damage to their natural environment. The promised prosperity has only percolated through to a tiny number of the forest dwellers.

A look at the list of the richest Malaysians clearly identifies the beneficiaries of the wealth of the rainforest. According to *Forbes*, the 77-year-old boss of Rimbunan Hijau, Tiong Hiew King, has assets of 1.5 billion US dollars; the fortune of the 74-year-old Samling founder, Yaw Teck Seng, and his sons, Yaw Chee Ming and Yaw Chee Siew, is estimated at 865 million US dollars; while that of the 62-year-old head of Ta Ann, Hamed Sepawi, a cousin of Taib, is put at 175 million US dollars.[20]

These estimates doubtlessly err on the conservative side and only consider the part of the timber barons' wealth that is disclosed. The man who made all these magnates rich through the concessions issued to them does not appear on the *Forbes* list at all. He is Taib in person, presumably the richest man in Malaysia. Being a politician exposed to public view, he keeps his wealth secret, but the Bruno Manser Fund estimates it to be at least 15 billion US dollars.[21]

WITH BIBLE AND CHAINSAW

What is the secret of the Sarawak timber barons' success? A look into the biography and background of Tiong Hiew King, the most successful businessman of his ilk, reveals a lot. Tiong was born in 1935 in Sibu, then still a small trading town on the lower reaches of the Rajang, the longest river in Sarawak. Like many of Sibu's inhabitants, Tiong's parents had emigrated from the port of Fuzhou (Foochow) in the south of China and moved to Sarawak early in the 20th century. Many of them had been Christians and welcomed the chance to leave China. The first thousand immigrants from Fuzhou arrived in Sibu in 1901, shortly after China's Boxer Uprising of 1900, when violent attacks on Chinese Christians were directed against anyone who had contacts with western missionaries.[22]

For Rajah Charles Brooke, the Chinese settlers were a godsend for the development of the fertile land on the Rejang. He refunded their travel expenses to Sarawak, gave each adult a hectare of land and lent them 30 Sarawak dollars each from the state budget, to be paid back in five years.[23] The Rajah made a good investment. Descendants of the Chinese immigrants to Sibu were later to become some of the most successful business people in Malaysia. Sarawak's Fuzhou community grew to 120,000 within a hundred years. However, Sibu business people now also play a leading role in forest destruction. Five of Sarawak's six biggest timber and plantation corporations are controlled by Fuzhou Chinese.[24]

Tiong Hiew King grew up in simple circumstances in a village downstream from Sibu, where his parents had a rice farm and a small rubber plantation. After completing the Methodist mission school, Tiong moved on to the Catholic Sacred Heart School before finding his first job in the timber business of his uncle, Wong Tuong Kwong, the founder of WTK.

In 1975, Tiong left his uncle's business and set up his own timber company, to which he gave the inspired name Rimbunan Hijau ("forever green").[25]

Then something happened to Tiong that marked him profoundly. The then head of the Sarawak government, Chief Minister Rahman Ya'kub (Taib's uncle), had him arrested and put in prison for several weeks. Rahman accused Tiong of communist sympathies, but perhaps he had simply become too rich for him and thus politically dangerous. Another timber businessman and leading opposition politician, James Wong, head of the Limbang Trading Corporation, was also arrested at the same time at Rahman's instigation under a similar pretext.[26]

After his release, to regain Rahman's political favour, Tiong grovelled in the most humiliating manner. When the chief minister flew to Taiwan to play golf there, Tiong followed along for the sole purpose of holding his umbrella for him on the golf course.[27] Tiong had learnt his lesson: Without politicians, nothing is going to function in the timber business. They are always in the stronger position in matters of concessions and permits. If necessary, they can even throw people into prison. Tiong learned the golden rule of the timber mafia: Stay in politicians' favour, share the spoils, and always show one's heartfelt gratitude.

When Taib became chief minister in 1981, Tiong immediately appointed Taib's favourite brother, Arip, to the board of Jaya Tiasa, the exchange-listed subsidiary of Rimbunan Hijau. Other family members and political friends of Taib were ushered into positions as directors and shareholders in twelve companies belonging to the Rimbunan Hijau group. Taib expressed his thanks to Tiong by issuing new logging concessions to him and by renewing existing ones for a total area of more than 1.5 million hectares[28]—worth more than 10 billion US dollars.[29] In this way, Tiong's company became one of Sarawak's largest timber corporations within only ten years of being set up.

It was more than mere attention to the political leadership that ensured the economic success of the timber businessman from Sibu. The strict work ethic practised by Tiong, who had had a protestant upbringing, also played its part. "We can probably ascribe Tiong's success in part to the religious values of his church," said the former pastor and human rights activist, Wong Meng Chuo, who, like Tiong, belonged to the Sibu Methodist church, and whose family he had known since childhood. "Work hard, don't waste time, practise thrift, and regard material wealth as a sign of divine mercy. That is typically Puritan, but also in line with the Chinese-Confucian tradition."

The protestant ethic and the spirit of capitalism was the title the German sociologist Max Weber gave to his famous 1904 treatise on the correlation between religion and economic success. A hundred years later, there could be no better model of Weber's view of the protestant capitalist than the timber baron Tiong Hiew King. Nor is he the only protestant among the timber barons. One of the sons of the Samling founder, Yaw Chee Weng, donated a plot of land in Bintulu for a new place of worship for the indigenous protestant church, Sidang Injil Borneo (SIB). As Joe Studwell's recent study *Asian Godfathers* showed, Protestantism also exerts an almost magical force on many other business magnates in Southeast Asia.[30]

"Like many Puritans, Tiong does not drink alcohol, does not smoke and does not gamble. This has allowed him to build up a good social image, and has been beneficial for his business. In addition, the church (and earlier the Christian missionaries) always formed an important commercial network," Pastor Wong explains. "But unfortunately, like many Christians, Tiong is very selective in his appreciation of religious precepts and ignores such values as social justice and protection of the environment."[31]

PAPUA CAMPAIGN

As the 1980s drew to a close, it was becoming clear that the aggressive exploitation of the Sarawak rainforests would very soon lead to a shortage of high-grade timber suitable for processing. The Malaysian timber barons started to look around for new reliable sources of timber. The nearest were the still untouched forests in the neighbouring countries of Southeast Asia, which promised a very high potential for exploitation. Indonesia remained closed to the Malaysians on political grounds. That meant concentrating activities on the virgin forests of Papua New Guinea, Cambodia, the Solomon Islands, and other Pacific island states. Thanks to the ample profits from deforestation in Sarawak, the timber corporations' war chests were full, ready for investment in new theatres of operation.

Sarawak's coat-of-arms depicts the rhinoceros bird, Guyana's the hoatzin, and Papua New Guinea's the bird of paradise. All these decorative armorial birds are inhabitants of the tropical rainforest and they all suffer directly from its exploitation. Thirty-nine species of birds of paradise live only in New Guinea and the rainforest regions bordering on it. Alfred Russel Wallace, the great explorer of the Southeast Asian archipelago, called them "Earth's most beautiful and extraordinary feathered inhabitants".[32] The males are known for their amazing courtship displays, and for centuries the indigenous peoples of Papua New Guinea have used their feathers for rituals and as decorations.

Like Sarawak and Guyana, New Guinea, the second largest island on Earth, with its extensive virgin rainforests, is one of the world's greatest centres of biodiversity. The fascinating creatures that live there include tree kangaroos, spiny anteaters, and giant snakes, not to mention the greenish-blue glistening Queen Alexandra's birdwing, the world's largest butterfly, with a wing span of up to 28 centimetres. In just the first decade of the 21st century, more than a thousand new animal species were discovered there.[33]

What is also unique is the cultural diversity of the population of New Guinea, where over a thousand indigenous languages are spoken. Politically, New Guinea is divided in two. West Papua (Irian Jaya) occupies the western half of the island, annexed by Indonesia in 1963; Papua New Guinea (which became independent in 1975, after being administered by Australia since 1919) is in the eastern half. Papua New Guinea's ethnic diversity is also a cause of political instability and, together with the island state's weak government structures, creates ideal preconditions for exploitation of its natural resources by external colonisers.

In 1989, Rimbunan Hijau was the first Malaysian timber corporation to begin harvesting Papua New Guinea's rainforests. One year later, an independent investigation by an Australian judge found that the country was suffering from disastrous management of its forest resources and massive corruption.[34] Tiong's company was shown to have systematically bought all the country's politicians who might have stood in the way of its interests.[35] It took Tiong only a few years to acquire a quasi-monopoly position in the country's timber industry. For his campaign into Papua New Guinea, the Sarawak Chinese made use of more than sixty companies, of which only some officially belonged to his Rimbunan Hijau group.[36] One particularly clever move was the creation in 1993 of a newspaper called *The National*, one of only two English-language newspapers in the country. This made it possible for the timber baron to exert political influence over the public as well—something he did with increasing success.

Tiong's competitor from Sarawak, the Samling boss, Yaw Teck Seng, also adopted exceptionally blatant tactics in Papua New Guinea. His company, Concord Pacific, started building a road from the remote village of Aiambak to the small town of Kiunga in the west of the country in 1995, purportedly as a "development project". It very soon emerged that the road built by Samling's sister company was just a pretext for felling trees in the previously intact rainforest. After a "construction period" of seven years, sixty kilometres of road had still not been built, but, instead, the most valuable logs had been stolen from thousands of hectares of virgin forest.[37] In 2011, the Papua New Guinea national court ordered the company to pay compensation totalling 97 million dollars to the indigenous inhabitants, who had been harmed by the project—a rare victory for the opponents of the timber corporations.[38]

A Greenpeace study established that in 2005, more than 80% of timber exports from Papua New Guinea were carried out by Malaysian timber companies. Two companies belonging to Tiong's empire exported 435,000 cubic metres of tropical timber within just five months, nearly half of the country's total timber exports.[39] Virtually all of this timber was exported in the form of logs to China and other Asian destinations, and the trickle-down of wealth in the country itself remained minimal.

The indigenous land owners—without whose permission no timber should have been felled—suffered the most from this theft.

Independent investigations have repeatedly shown that Rimbunan Hijau has been involved in corruption and illegal logging since the very beginning of its activities in Papua New Guinea. The World Bank estimated in 2006 that 70% of the trees felled in Papua New Guinea were cut down illegally.[40] Since, however, the country's leading politicians and its judicial system did nothing against the Malaysian economic colonisers, there were never any consequences. On the contrary: Tiong and his companies have had the red carpet laid out for them by the country's corrupt elite, who also earn a cut from the forest destruction.[41]

Worse still, acting on a proposal from the government of Papua New Guinea, in 2009, Queen Elizabeth II, who is still formally the country's titular head of state, appointed Tiong Hiew King an honorary CBE—Knight Commander of the Most Excellent Order of the British Empire for his services in Papua New Guinea.[42]

Today, Tiong Hiew King, the son of an immigrant who settled in Sibu, finds himself at the head of an intricate conglomerate worth several billion US dollars. In addition to timber and plantation operations in countries such as Gabon, Equatorial Guinea, Indonesia, New Zealand, and Russia, it also runs media outlets in Hong Kong, Malaysia, Canada, and the USA; real estate in Australia, China, and Papua New Guinea; and oil and gas operations in Myanmar and Malaysia.[43] The fact is that the initial capital injection for all these later expansions to his businesses came originally from the logging concessions granted to him by Taib and from the decimation of the Sarawak rainforest.

OUT OF AFRICA

Hidden from scrutiny by world public opinion, the Malaysian timber industry became one of the biggest players in the tropical forests of Central and Western Africa. Since the 1990s, the Malaysians have been active in particular in Gabon, Equatorial Guinea, and Congo-Brazzaville.[44] Here too the companies from Sarawak are among the front runners, especially Rimbunan Hijau, operating through its African subsidiary, Shimmer International. It plays a key role in supplying China with African timber. According to the International Union for the Conservation of Nature, Rimbunan Hijau is a key supplier of logs of the tropical okoumé tree, which is much in demand for the production of veneers in China.[45]

In expanding into Africa, the company owned by the Protestant Tiong showed no compunction about rubbing shoulders with despots of the most repugnant sort, such as Teodoro Obiang Nguema, who rules over oil-producing Equatorial Guinea. Obiang, whose regime has been accused of systematic torture, murder of its opponents, and numerous other violations of human rights, is regarded as one of the worst dictators in Africa.[46] His son, "Teodorin", whom his father appointed as minister of forests, demanded the payment of enormous bribes from foreign investors as the price for entering the timber business.[47] The Malaysians did not see that as any reason for not going ahead. In the mid-1990s, they started doing business with the dictator's son, and it did not take them long to establish a quasi-monopoly in the country. According to investigations carried out by the US Department of Justice, Rimbunan Hijau paid "Teodorin" a bribe of 30,000 Central African Francs (roughly 60 US dollars) for every cubic metre of timber harvested. In exchange, the group was exempted from compliance with the forestry laws and allowed to operate in protected national parks.[48] Most of the trees were felled in only a few years, and the country's timber exports soon collapsed from 700,000 cubic metres in 1997 to 300,000 cubic metres in 2004.[49] Well over a hundred million dollars in bribes must have made their way into the pockets of the forestry minister over that time.

In the neighbouring west African country of Gabon, ruled for more than forty years by the dictator Omar Bongo—who died in 2009—, Rimbunan Hijau was one of the biggest timber companies. Here the Malaysians have control over 1.69 million hectares of tropical forest, about 17% of the country's exploitable forestry land.[50] Most of the trees felled by Rimbunan Hijau up to 2010 (several million cubic metres of timber since 1996) were transported unprocessed to China, so that very little added value was obtained in Gabon itself.[51] This was the same pattern as in Congo-Brazzaville, from which the Malaysians have been exporting timber to China since 2001. The export volume in 2004 was 500,000 cubic metres.[52]

CRIMINAL TOURISM INVADES KHMER COUNTRY

When Credit Suisse presented the Samling Group to the international investor community in Hong Kong for its 2007 stock market launch, one important episode in the timber company's history was omitted from the grandiose portrait of the company.[53] Between 1994 and 2001, Samling had been the biggest timber corporation active in Cambodia. After that, the Malaysians had been thrown out of the Land of the Khmer by government decree. The Cambodian government terminated all logging in the country as of New Year 2002. This decision resulted in large part from the manner in which the Yaw family from Sarawak had been conducting its business there.

Yaw Teck Seng, a Sarawak Chinese with origins in Guangzhou (Canton), South China, was born in 1938. He entered the logging trade as a 25-year-old in 1963, the year of Sarawak's independence. In 1976, he founded Samling (which means "three trees" in Mandarin). It was initially as large as Rimbunan Hijau, the company belonging to his rival, Tiong, from Sibu. At first, Yaw was as successful as Tiong in obtaining logging concessions from Chief Minister Taib, and had accumulated a total of 1.6 million hectares by the mid-1990s.[54] After that, Tiong's business acumen proved superior—after substantial investments abroad, Rimbunan Hijau became much bigger than Samling.

Three years after entering Guyana, the Yaws took a second important step into expansion abroad, closer to home in nearby Cambodia. On 18 August 1994, the Samling subsidiary, SL International, received logging concessions for 787,000 hectares of forest from the Cambodian government (12% of the country's forests), making it overnight into the biggest player on the Cambodian timber market.[55]

A year before, after decades of civil war and dictatorship, the country on the Mekong had elected a new government under the supervision of the United Nations. The wounds caused by the Khmer Rouge genocide, in which two million Cambodians were killed in the 1970s, were still fresh. In such a fragile post-conflict situation, the government institutions were still very weak, a situation shamelessly exploited by Samling.

The British NGO Global Witness began asking questions about the legality of Samling's activities in Cambodia, given that the huge concessions had been awarded by the government without any prior consultation of parliament. As in Guyana, the Yaw family group had also obtained tax exemption for several years in exchange for its "investments" in Cambodia.[56] The human rights organisation discovered that the Malaysians had purchased timber illegally from the Cambodian army and border police, that they had paid protection money to the army and the Khmer Rouge and that they had prevented people living in the concession areas from cutting down trees for their own use.[57]

In 1996, massive illegal logging in a game reserve and the violation of other legal provisions by Samling led the Cambodian government to impose a temporary ban on any further tree felling in the country. The Malaysian "investors" were obviously far from pleased with this and chose simply to ignore the government decree. In April 1997, Han Chen Kong, the director of the Samling subsidiary in Cambodia, received a written warning from the minister of agriculture, accusing his company of illegal logging.[58]

Three years later, the Asian Development Bank examined Cambodian forestry policy and found a "total failure of the system". The biggest concession holder was a Samling subsidiary, SL International, which was heavily implicated in this devastating judgment.[59]

So it came as no surprise that Samling was forced out of Cambodia two years later. It is not known how many trees were felled and exported by Malaysian loggers on behalf of the Yaw family during its six-and-a-half year reign in Cambodia. Samling itself, however, is obviously at pains to expunge this dark chapter from the company history. It was only under public pressure that Samling admitted on its website in May 2007 that an "indirect subsidiary company", SL International, had been active in Cambodia until the end of 2001, but it denied "to the best of its knowledge and belief" all the accusations of illegal logging.[60]

Unfortunately, Samling's experience in Cambodia did not teach the company the importance of playing by the rules. The profits to be made from illegal logging were simply too great. Take, for example, a 2009 newspaper report from the Pacific Solomon Islands, in which a village chief, Henry Stardora, called on a Malaysian logging company, operating under the name "Samling Sun", to leave the land of the indigenous people of Hailadami: "Mr Stardora said Samling Sun was a logging company that typically disobeyed the laws of the land. He said that the company was also active in other provinces of the country, and the owners of the land would have to be wary of it."[61]

The Yaws' Samling corporation and its associated companies have systematically violated forestry laws throughout their history. Fines in Guyana, a multi-million dollar penalty in Papua New Guinea, illegal logging on the Solomon Islands—the business conducted by the Malaysian loggers is both corrupt and environmentally destructive. Yet it is only very rarely that the perpetrators and profiteers are called to account.

CRONYISM IN AUSTRALIA

Ten days before Christmas 2011, in the heat of the Australian summer, a dark-haired young woman climbed a 400-year-old eucalyptus on Mount Mueller on the island of Tasmania, Australia's most southerly state. The 30-year-old grammar school teacher, Miranda Gibson, then ensconced herself on a makeshift platform 60 metres above the ground, defying both the elements and Australia's forestry policy.

"I'm not going to leave this tree until this forest is protected," Gibson pledged, after she had climbed her eucalyptus, which she had declared to be the "Observer Tree" overlooking the threatened virgin forest. Four months previously, the Australian prime minister, Julia Gillard, had announced that more than 400,000 hectares of virgin rainforest in southern

Tasmania were to be placed under protection. However, this announcement was not followed by any sort of action, and when the lumberjacks advanced into territory close to Mount Mueller, Gibson felt the time had come for action and decided to occupy one of the oldest trees in the region. She spent more than a year living in the treetop, thereby establishing a new Australian record in "tree-sitting". While up there, she faced scorching heat, stormy winds, and even ice and snow, without once leaving her tree.

Thanks to a laptop and wireless internet, it was possible for the environmental activist to follow developments in forest policy from her treetop home and to comment on them every day in her blog.[62] She called on the visitors to her website to "help me show the world that it is time for genuine forest protection". Her extraordinary action, backed by the local NGOs Markets for Change, Huon Valley Environment Centre, and The Last Stand, attracted a media blitz. In addition to Australian media, reports on her action to protect the Tasmanian forests were carried in newspapers such as the British *Guardian* and on television networks like the *BBC*, *CNN*, and *Al Jazeera*.

From her vantage point in the forest canopy, Miranda Gibson looked down on the impressive Styx Valley (named after the River of Death in Greek mythology) with its mighty tree ferns and giant eucalyptus (*Eucalyptus regnans*), the tallest deciduous trees on Earth, growing to a height of up to a hundred metres. In addition to a wide variety of endemic plants, Tasmania's wet eucalyptus forests are home to endangered species, such as the quoll and the Tasmanian devil, a carnivorous marsupial the size of a dog, which became extinct in mainland Australia a long time ago. It is also the breeding ground of the distinctive white goshawk, the wedge-tailed eagle and the colourful swift parrot, a species of parrot that nests only in hollows found in blue gum trees (*Eucalyptus globulus*). Further up Mount Mueller is the boundary of the UNESCO-protected Tasmanian Wilderness World Heritage Area, containing three large national parks. Environmentalists, however, claim that the park borders were deliberately drawn to facilitate industrial logging in the extensive virgin eucalyptus forests.

As in so many places around the globe, the Malaysian logging industry also has its fingers in the pie here. The Ta Ann timber corporation,
which has close ties with the Taib family, plays a central role in the
destruction of Tasmania's natural forests. When the Australian federal
government and the Tasmanian state government agreed on the protection of 430,000 hectares of Tasmanian forest of particular biological
value in August 2011, Ta Ann made sure it had a guarantee for the annual
delivery of 265,000 cubic metres of timber for the manufacture of veneers. It soon emerged that it was only possible to supply such quantities
by chopping down the ancient eucalyptus forests, making nonsense of the
supposedly historic deal.

Of all the Malaysian timber corporations, Ta Ann has the closest
ties with the Sarawak chief minister. The company was founded in the
mid-1980s by Taib's cousin, Hamed Sepawi; the PBB politician Wahab
Dolah (the PBB party forms Taib's principal political power base); and
the businessman Wong Kuo Hea. It was immediately granted huge logging concessions in Sarawak—and, of course, there was no such thing as
a public call for open bidding.[63] As chairman and principal shareholder,
Taib's cousin, Sepawi, is the strong man at Ta Ann.[64] Insiders, however,
suspect that Sepawi is merely a front man for Taib himself, who controls
the corporation from behind the scenes.[65]

Hamed Sepawi, who was born in 1949, is one of the closest confidants
of his cousin, who is twelve years older. In Malaysia, he has significant
holdings in at least fifty-five businesses. He is chairman and a large shareholder of the construction group Naim Holdings (16%); he is a major
shareholder of Samling; and is chairman of Sarawak Energy, the state-
owned power utility. Until 2006, he also presided over the state-owned
Sarawak Timber Industry Development Corporation, before Taib took
over that office in person. Sepawi also has business interests in the USA
and Australia, where he is active with Ta Ann not only in the processing
of wood, but also in real estate and as a financial investor.[66]

In Sarawak, Ta Ann has not only cut trees in the Penan's rainforest,
but has also destroyed the habitat of orang-utans in the interior of the
country, in a region known as the "Heart of Borneo", the protection of

which had been agreed in a joint commitment by the governments of Malaysia, Indonesia, and Brunei.[67] According to the company's own figures, Ta Ann today holds 362,000 hectares of logging concessions in Sarawak, not to mention its 313,000 hectares of plantation licences.[68]

In looking for new sources of timber to supply the Japanese market, the principal buyer of its products, Sepawi's corporation expanded into Australia in 2006—and received subsidies running into millions from the Tasmanian government for building two sawmills to produce veneers from eucalyptus trees. By concluding supply contracts with Forestry Tasmania, a company owned by the state of Tasmania, Ta Ann signed contracts to keep timber supplies flowing for the next twenty years. What makes this an interesting business for Ta Ann is that Tasmanian timber costs only about one quarter as much as Malaysian hardwood. Despite the low timber costs, Ta Ann has only reported losses since starting up in business in Tasmania and has thus never paid business taxes—just like Samling in Guyana.[69] The profits were made elsewhere. In 2012, Ta Ann reported sales of 29 million Australian dollars and a loss of 10 million dollars in its Tasmania operations. On the other hand, the Malaysian company managed to obtain generous government subsidies from the Australian government—to the tune of 45 million dollars since the company's arrival in Tasmania.[70]

"Ta Ann's operations in Australia are neither economically nor ecologically sustainable," says Jenny Weber of the Huon Valley Environment Centre, a small Tasmanian environmental organisation, which is committed to protection of the wilderness in Australia's southernmost region. "Ta Ann has brought Taib's disdain for the environment with it to Tasmania. The corporation has now become one of the driving forces behind the large-scale felling of our ancient rainforest, which ought to be protected."

Despite Ta Ann's losses, the Tasmanian government has been able to profit directly from business with Hamed Sepawi. The state-owned electricity company, Hydro Tasmania, has received lucrative consultancy jobs and contracts from Sarawak Energy for dams planned in the Sarawak rainforest. Hydro Tasmania only withdrew at the end of 2012, when a Bruno Manser Fund supported delegation travelled to Australia to report on the devastating consequences of these projects for the indigenous peoples of Sarawak.[71]

Victory came to the tree-squatting Miranda Gibson on 31 January 2013, after more than a year at the top of her eucalyptus. The Australian government announced that it would apply to UNESCO for an extension of 170,000 hectares to the Tasmanian Wilderness world-natural-heritage reserve.[72] On 7 March 2013, a happy and much relieved Miranda Gibson left the canopy of the Tasmanian rainforest after 449 days.[73] Three months later, the UNESCO World Heritage committee approved the request from the Australian government. The area around the "Observer Tree" and other virgin forests nearby became part of the world's natural heritage.[74] Without the extraordinary and crucial commitment of people like Miranda Gibson and Jenny Weber, preservation of Australia's forests would not have been possible.

The battle for Tasmania's forests, however, is far from over. When a conservative government came to power in September 2013, Australia's new Prime Minster, Tony Abbott, applied to UNESCO to delist 74,000 hectares of Tasmanian natural forests from the World Heritage site.[75] Fortunately, UNESCO rejected the request.

CORE BUSINESS: ENVIRONMENTAL DESTRUCTION

The cases described above are just a few examples of the central role played by Malaysian logging companies in the destruction of rainforests around the globe. Over the past two decades, logging companies from Malaysia have also been active in Surinam, Brazil, Belize, Cameroon, Indonesia, Russia, New Zealand, the Solomon Islands, and Vanuatu. Members of the Taib family or other leading Malaysian politicians have been directly involved in many of these companies and have taken their share of profits from worldwide rainforest clearance.

These timber corporations (the majority of them from Sarawak) thrived in a country that not only never committed itself to the sustainable harvesting of timber but also exported their culture of corruption and environmental destruction to developing countries with weak institutions. In that way, they have been able to maximise harvests, exporting the unprocessed logs to Asia. The countries of origin of the tropical timber have received hardly any sustainable economic benefits from the removal of their trees by the magnates from Malaysia, and the price has been social and environmental damage.

As long ago as 1997, a Greenpeace report urgently warned the Brazilian parliament about the advance by Malaysian logging companies into the Amazon Basin, the world's largest virgin forest region.[76] "Our experience has shown that Asian logging companies are among the leading destroyers of tropical rainforests," Greenpeace wrote at the time and emphasised the key role of the corporations from Sarawak.[77] Other NGOs stressed that the expansion of the Malaysian logging companies was being actively pushed by the Malaysian government.[78] At that point in time, Samling and WTK had just acquired large concessions in the Amazon. However, as a result of political resistance inside the country, these Brazilian logging operations failed to take off and the Malaysians were soon forced to withdraw.[79]

Twenty years after their international expansion, the timber corporations from Sarawak are still playing an important role in the worldwide tropical timber business, which is far from becoming sustainable. During this time, all of these corporations have not only grown, they have diversified. One thing they all have in common is that they have thrived thanks to a political climate rife with corruption and to an extremely lax application of the forestry laws, making it possible to flout rules more or less at will. As a result of the high profit margins in tropical timber, they subsequently moved into new fields of business, such as hotels, real estate, media, shipping, and palm oil production. Making profits at the expense of the natural environment of the tropics has, however, remained their core business.

09

GREEN WASTELAND

As logging has not proved sustainable, more and more of Sarawak is being covered by sprawling oil-palm monocultures. The palm oil industry has now become the principal cause of destruction of the rainforests in Southeast Asia, but new projects to construct dams are threatening the livelihood of tens of thousands of indigenous people too.

HIGHWAY OF FEAR

Two roads lead from the oil-drilling town of Miri on the north coast of Borneo to Bintulu, the third largest town in Sarawak, which is almost 200 kilometres away. The older of these two roads takes a route through the hilly terrain of the country's interior, past various longhouses occupied by indigenous Iban. The second, newer, road clings more closely to the coast.

The coastal road passes Luak Bay to the south of Miri, with its "millionaire mile", where Sarawak's timber barons have built pretentious palaces and had high fences and cast-iron gates erected to protect themselves from the enquiring gazes of the public. Large "Keep out!" signs make it unmistakably clear what lies in store for trespassers, with outlines of security men with firearms aiming at them.

After a while, palms begin to line both sides of the road, and beyond the small town of Niah the countryside becomes a moonscape, extending for dozens of kilometres. The endless rows of oil palms on both sides of the road stretch to the horizon through the eerily quiet countryside devoid of any sign of wildlife or human presence. On the roadside, there are warnings not to travel after nightfall on account of gangs of armed robbers. This section of the coastal road is also known as the "Highway of Fear".[1]

On recently cleared plots of land between the plantations, the peaty earth laid bare is a poignant reminder that a lush lowland rainforest used to grow here until only a few years ago. That is now a thing of the past. In all directions, the gigantic monocultures sprawl as far as the eye can see. There is no sign of any animals, or of any other trees apart from oil palms.

Elaeis guineensis is the scientific name for the oil palm, which is indigenous to West Africa and was first planted in Southeast Asia in the 19th century. The advance of the oil palm as a plantation species only began in the second half of the 20th century, first in Peninsular Malaysia, and then in Indonesia, followed by Sabah and Sarawak—and then all around the world, everywhere where tropical rainforest once grew. Taking just Malaysia, the surface area covered by oil palm plantations increased by more than half to 4.2 million hectares between 1990 and 2005, equal to over one eighth of the country's total area.[2]

Seen in purely commercial terms, the growing of oil palms has been a boon for Malaysia's national economy, but it has meant calamity for the rainforest—and for the country's indigenous communities. In Southeast Asia, the expansion of oil palm plantations is today the principal cause of rapidly advancing deforestation. A team of scientists from the United Kingdom investigated the effects of the oil palm boom on biodiversity and came up with alarming findings. Most of the new oil palm plantations created in Malaysia and Indonesia appear at expense of rainforest; they lead directly to a dramatic decline in the number of natural species. The plantations contain less than 15% of the biodiversity of the rainforest and provide a habitat mainly for a few ubiquitous varieties, whereas endangered species of rainforest fauna and flora disappear almost in their entirety.[3]

There are good reasons why no other cultivated tropical plant has been planted in such immense numbers in recent decades as the African oil palm. Firstly, there is its high yield per unit of land occupied, and secondly, there are the multiple uses to which it can be put. It is possible to harvest the reddish fruits just three to four years after planting the young palms. Palm oil, which has a high energy content, is pressed from the flesh of this fruit; palm kernel oil, which has a yet higher fat content, is extracted from the kernels, and pellets are produced from the dried shells of the seeds.

If not for the destruction wreaked on tropical forest, it might be easy to wax lyrical over the rich fruit. The food, cosmetic, and chemical industries are all very keen to lay their hands on it as a raw material, and it is estimated that one out of ten products on the supermarket shelf today contains palm oil. We consume our first dose of palm oil at breakfast without noticing it, because it is often an ingredient in margarine or cereal or the mixed spices used to flavour fried eggs. Given its high oleic content, palm oil is one of the few vegetable oils that is solid at room temperature, which greatly facilitates its processing. Furthermore, palm oil does not contain any of the harmful trans-fatty acids that have been targeted by the health authorities in recent years. Finally, it is comparatively cheap.[4]

With an annual worldwide production of 57 million tonnes (2013), palm oil is today the most important vegetable oil. It has a global market share of over one third (35%), placing it ahead of soybean oil (27%), rape-seed oil (15%), and sunflower oil (9%).[5] At the time of writing, Indonesia and Malaysia together produce more than 85% of the world's palm oil, but other tropical countries are rapidly catching them up—often as a result of the foreign activities of Malaysian palm oil corporations. Between 2009 and 2013, international investors announced the planting of more than 1.6 million hectares of oil palms in the rainforests of the Congo Basin in Africa. Various Malaysian palm oil companies are at the forefront of these operations, including Sime Darby, Atama Plantations, and Felda.[6] The key reasons for the continuing palm oil boom are the production of biofuels from vegetable oils, and an increasing need for vegetable oils in the production of foods.[7]

For indigenous communities, such as the Iban people living in the Sarawak lowland, the planting of oil palms on an industrial scale usually goes hand in hand with the loss of their traditional land. It is often not until the bulldozers roll up that they realise that the Taib regime has secretly given their land away to oil palm corporations without seeking their consent. More than 90% of the oil palms in Sarawak have been planted in huge plantations by big companies and the state. Given that, once picked, the oil palm fruits start to rot very quickly, it is essential to process them within 24 hours of harvesting, which calls for sophisticated logistics. The short shelf-life of the fruit favours large companies that are able to achieve much higher returns with the cultivation of oil palms than small-scale farmers can. For the operation of a palm oil mill to be profitable, there must be at least 4,000 hectares of oil palms within its catchment area.[8]

SUSTAINABLE PALM OIL?

An example of what palm oil corporations—with Taib's complicity—are capable of was witnessed first hand by the inhabitants of Long Teran Kanan, a small indigenous community on the Tinjar River in the north of Sarawak. In the mid-1990s, Taib's ministry of agriculture granted plantation licences covering 7,800 hectares of land on communal territory to the Rinwood Pelita Company, a joint venture between a private business (Rinwood) and the Taib-controlled State Development Agency, Pelita, without any prior approval by the village inhabitants. Not long afterwards, the bulldozers turned up and began flattening the secondary forest and a large area of arable farmland belonging to the village, and preparing the land for planting with oil palms. They not only felled all the forest trees but also cleared cultivated pepper plants, cocoa shrubs, and durian trees. The drinking water supply was contaminated, and even rice fields fell victim to the bulldozers.[9]

"All my plants were carted off by Rinwood's bulldozers," one of the village inhabitants, Baya Sigah, recalls. "I used to have about 3,000 cocoa shrubs, and they all went. There was no genuine consultation. Sometimes, the bulldozers turned up on Sundays, when the people were in church. Some people were compensated, others not. I, for instance, have never received any compensation."[10]

When the village inhabitants tried to stand up to the tyranny, the palm oil company sent in gangs of thugs to intimidate them. Neither the police nor the local authorities did anything to protect the village inhabitants, and, instead, took the company's side. In 1997, four of the village inhabitants consulted the indigenous lawyer, Harrison Ngau, and initiated court action against Rinwood and the Sarawak government on behalf of their community. They demanded recognition of their land rights, cancellation of the Rinwood/Pelita plantation licence, and a restraining order against the company's employees from entering the community's land.

Twelve long years elapsed before a local court decided on the issue and awarded rights over 3,279 hectares of land to the community. The court, however, found that, as the oil palms had been planted in the meantime, it would suffice for the villages to receive compensation for use of their land. Even this decision went too far for the defeated palm oil company and the Sarawak government, and they immediately lodged an appeal against it.

While all this was going on, Rinwood had been bought up by the Malaysian palm oil giant, IOI—one of the biggest corporations in Southeast

Asia, which also owns refineries in the Netherlands, the USA, and Canada. The IOI founder, Lee Shin Cheng, appears in fourth place of the Forbes list of the richest Malaysians, with assets estimated at 5.2 billion US dollars.[11]

IOI is also a founding member of the Roundtable on Sustainable Palm Oil (RSPO), an association under Swiss law set up in 2004 by the WWF, various retail businesses, and the palm oil industry. It has its registered office in Zurich. A representative of IOI sits on the executive board of the RSPO, which has its main office in Kuala Lumpur.[12] On paper, the RSPO's sustainability criteria sound pretty good. They would, for example, prohibit the creation of oil palm plantations on native land without the consent of the indigenous communities affected. In practice, however, the RSPO criteria are paper tigers, and the administration of its label is an inefficient, bureaucratic piece of machinery powerless to force the powerful corporations that finance the label to adhere to its principles.

In the case of Long Teran Kanan, the acquisition of Rinwood by IOI with its "sustainable" production resulted in the immediate destruction by the new owners of several hundred hectares more of secondary forest and fruit trees and their replacement by young oil palms. At the end of 2009, a team of reporters from the BBC visited the region and filmed appalling footage of the large-scale destruction of the vegetation. IOI defended itself with the argument that it had spoken with the people in the village, and that the RSPO had certified that it was managing land in Sarawak responsibly.[13] That is what the creation of a "sustainable" palm oil plantation looks like in the real world.

At the same time, it became known that IOI had illegally set up huge new oil palm plantations in the Indonesian part of Borneo. It had cleared hundreds of hectares of rainforest in the Ketapang district in West Kalimantan and had drained swamps without the necessary permits, and without having clarified native customary rights. Falsified documents had been submitted to the Indonesian authorities, in which IOI maintained that nothing was going to be done with the newly acquired land until environmental impact assessments were available. That was, however, patently false. The scam was uncovered when a Dutch environmental organisation, Milieudefensie (Friends of the Earth), published a study on IOI in Kalimantan in March 2010.[14]

Despite international protests and lengthy negotiations between the indigenous inhabitants of Borneo and IOI, exposing these fraudulent practices has had no real consequences for the palm oil corporation. It is true that the RSPO temporarily suspended issuing of certificates for additional IOI plantations, but the company is still allowed to present its palm oil on the market as "sustainably produced".[15] It is to be expected that the negotiations between IOI, Long Teran Kanan, and the Malaysian courts will drag on for many years to come. In Indonesia, the indigenous communities have given up. IOI palm oil from Ketapang is likely to appear on the market in just a few years as a "sustainably produced" product.

The invention of the RSPO sustainability label is a godsend for the palm oil industry. Wherever it faces criticism, it can simply invoke the argument that it produces "sustainably", without needing to make any substantive changes to its business practice. The label is particularly useful for opening up the market for new and controversial fields of business, such as the use of palm oil as a fuel. Nearly all the big palm oil producers have noticed this, and are now selling RSPO-certified palm oil to Western Europe and other industrialised countries. As far as the rest of their palm oil is concerned, they will always find a sufficient number of purchasers in markets where consumers are less demanding.

In 2013, 108 companies and organisations from Malaysia were registered as RSPO members, including the palm oil giants IOI and Sime Darby. Even the Taib family has now joined in the business of "sustainable palm oil". Through his company, Achi Jaya Plantations, Taib's brother Onn possesses 12,000 hectares of RSPO-certified oil palm plantations in Johor, West Malaysia, which he bought in 2004 for 512 million ringgits (159 million US dollars). In April 2013, Onn reportedly negotiated with IOI for a sale of his plantations for a much higher price but the deal fell through.[16]

What is the true value of the RSPO for the environment and the indigenous peoples of Borneo? On the one hand, it is to be welcomed that environmental and social standards have been defined for palm oil production and that all the RSPO members have undertaken to abide by them, at least on paper. On the other hand, these criteria are weak—especially concerning environmental protection—and compliance is extremely difficult to monitor and enforce. The IOI case shows that a label is not worth much when the business interests of the palm oil industry are at stake, and, moreover, that the RSPO as an organisation is not sufficiently independent of the industry.

The WWF has not furnished any response regarding the impact of the RSPO either, although it was one of the participants in the founding of "sustainable palm oil". It is true that the WWF has published a study showing the potential economic benefits of RSPO certification for palm oil producers, but a decade after the RSPO was set up, there is still no evaluation of its impact on rural populations and the environment.[17]

It is doubtful that a product can really be called sustainable if it has been mass-produced in extensive monocultures at the expense of the rainforest. Rainforest could grow on every square metre of oil palm plantations. Scientists at Princeton University and at the Swiss Federal Institute of Technology in Zurich have calculated that between 1990 and 2005, over 4 million hectares were deforested in Malaysia and Indonesia just to make way for oil palms.[18] The victims of the capital-intensive palm oil industry are not only the indigenous peoples living off the rainforest, but more and more of the threatened species of fauna and flora of the tropics, of which the most spectacular include orang-utans, forest elephants, and the world's largest flowering plants (rafflesia and titan arum).

Up until now, no evidence has been forthcoming to show that the RSPO label has protected any rainforest at all. It is clear, however, that the RSPO has made the controversial palm oil more respectable, thereby opening up new markets. To take the example of Switzerland, the annual consumption of palm oil grew from 21,000 tonnes in 2004, when RSPO was set up, to 34,000 tonnes in 2012, an increase of 62%.[19] In the European Union, sales of palm oil accelerated when, at the end of 2012, the European Commission recognised the RSPO label as sufficient evidence of sustainability for palm oil to be allowed access to the subsidised market for fuels from renewable sources.[20]

One example of this is the Neste Oil company, which is controlled by the Finnish state and is a major customer of IOI. The company justifies the production of fuels from palm oil with the argument that from 2015 onwards, only palm oil certified as "sustainable" will be used.[21] With its two new refineries in Singapore and Rotterdam, which cost Neste Oil 1.2 billion euros (1.6 billion US dollars), the company has an annual production capacity of 1.6 million tonnes of agri-diesel available to it.[22] In 2012 alone, Neste Oil processed 1.36 million tonnes of palm oil for fuel production, corresponding to 2.7% of worldwide palm oil production.[23] The state-owned Finnish company has become one of the world's largest consumers of palm oil.

Norway has found a totally different answer from Finland. Between 2010 and 2012, the Norwegian state pension fund divested all its shares in IOI and another 29 palm oil corporations from Malaysia, Indonesia, and Singapore that are responsible for clearing tropical rainforests. It is a decision that carries weight, since the Norwegian pension fund, which manages assets worth more than 650 billion US dollars, is the world's biggest institutional investor. The Norwegians have also sold—among other investments—all the shares they once owned in Ta Ann Holdings.[24]

STATE-PLANNED LAND GRABS

In Sarawak, the rapid expansion of palm oil plantations is intimately linked to the corruption of Chief Minister Taib. It was certainly no matter of chance that Taib, in addition to his office as head of the state government, also grabbed the ministry of resource planning for himself. There are no publicly available maps showing the land areas covered by palm oil licences and logging concessions, and the government's land transactions are amongst the best-guarded state secrets. After all, pinching land from the state and then distributing it amongst his supporters are central pillars of Taib's power. It is only thanks to courageous insiders within the Sarawak government, who feel they cannot permit Taib's

appropriation of land and the destruction of Borneo's rainforest to go unnoticed, that information occasionally seeps out to the public.

The clandestine meeting with one such informer took place at the end
of 2010 in a small family-run hotel in the Indian quarter of Singapore. For
his personal safety, I was not permitted to enter into any sort of contact
with him in Malaysia, nor was I permitted to telephone him either before
or after our meeting. He was much too afraid of being spied on by the
Special Branch, Malaysia's political police, and of Taib's revenge for the
divulgence of one of his biggest secrets, which was handed to me on an
inconspicuous grey memory stick.

The file with the apparently innocuous name of "land and survey"
contained a list of names, dates, prices, and surface areas, and appeared at
first to be of little interest. On closer perusal, however, the long list from
Sarawak's land and survey department turned out to be almost too hot to
handle. Line for line, it established how, between 1999 and 2010, Taib
had distributed 1.5 million hectares of state-owned land worth several billion dollars to members of his family and his friends as well as to business
associates and politicians close to him.[25] They were granted exclusive
rights for a period of up to sixty years to harvest all the timber and to plant
oil palms or other monocultures on land belonging to the state and the indigenous peoples. There was no public scrutiny of the distribution of
these lands. Taib made the decisions himself and gave no public account.

At least 199,000 hectares of land for oil palms went directly to Taib's
closest relatives and to companies in which members of his family have
official holdings. Taib's son Abu Bekir; his sister Raziah; his cousin
Hamed Sepawi; his niece Elia Geneid; and the Delta Padi company, in
which Taib himself holds shares, are amongst those favoured with most
land, the value of which is estimated to run into several hundred million
dollars. More than 45,000 hectares of this land was placed in the hands
of Taib's family for free, while the price paid for the rest of the land was
way below its market value.[26] Much of the land was resold within a short
period of time at a very much higher price. On her website, *Sarawak Report*, the journalist Clare Rewcastle documents one case in which Taib's
brothers Arip and Ali pocketed a profit of 40 million ringgits (approxi-

mately 12 million US dollars) within twelve days in May 2004 from re-selling 15,000 hectares of state land.[27]

Global Witness wanted to find out more about the mechanisms used by Taib for grabbing land and sent an experienced investigator to Sarawak in 2012. He presented himself at the land and survey department, claiming to be the representative of foreign investors wanting to acquire land for oil palm plantations in Malaysia. He carried at all times a tiny hidden camera and secretly filmed his meetings with Taib's agents.[28]

Shortly after Andy S. (the investigator's pseudonym) had established contact with the land and survey department, an official there sent him straight on to Taib's cousins Fatimah (48) and Norlia (55), two daughters of his uncle and predecessor, Rahman Ya'kub. After a feud lasting many years, Taib had publicly patched things up with Rahman at the beginning of 2008 (see chapter 5). Now that peace had been established within the family, it had become possible for Rahman and his eight daughters to resume business with the chief minister.

The investigator met Rahman's daughters for sales negotiations in the Pullman Kuching, one of the best five-star hotels in Sarawak. In the apparent seclusion of the hotel room, Fatimah and Norlia dropped their inhibitions as they explained how land grabbing worked. They used as their example Ample Agro company—owned by Fatimah and Norlia along with five of their sisters, including Norah (who sits for Taib's party, PBB, in the Malaysian federal parliament) and Khadijah, who is married to a brother of the Malaysian prime minister, Najib Razak.[29]

Early in 2011, Ample Agro had received a concession from Taib to clear 5,000 hectares of rainforest in the Tekoyong district and to use the land as an oil palm plantation until 2071. It had paid roughly 330,000 US dollars for the rights and had also agreed to an annual lease at roughly one dollar per hectare. Although that part of the forest had been used by indigenous Iban communities for more than a hundred years, the Taib government denied them any rights at all over the forest, which had been classified as state-owned land.

Fatimah, Norlia, and the Ample Agro company offered to sell this land concession to the Global Witness investigator for 49.4 million ringgits (some 15 million US dollars). Without doing a stroke of work, Taib's cousins were hoping to pocket a vast profit—with no qualms of conscience and in major violation of the rights of the indigenous inhabitants. "These people are squatters on the land," Norlia told the hidden camera. "They have no claim to the land, but they can make life difficult for you if they are not kept quiet."

Rahman's daughters were not the only ones to offer land to Andy S. He received a further offer of 32,000 hectares from Billion Ventures, a company owned by Taib's confidant Hii Yii Peng. The timber baron wanted 230 million ringgits (around 70 million US dollars) for it, and had entrusted his nephew with the negotiations. The latter was frank with the investigator (not knowing, of course, that he was being filmed). Ten per cent of the purchase price would have to be handed over directly to the chief minister, he said, but his uncle Hii Yii Peng would take charge of that. If that deal had materialised, Taib would have pocketed a kickback of some seven million US dollars.[30]

The Billion Ventures land, a magnificent secondary forest, is located in the north of Sarawak, in the upper reaches of the Limbang River, near the Kelabit village of Long Napir, only a few kilometres from the UNESCO-protected Mulu National Park. It has been used by indigenous Penan and Kelabit communities for centuries. Taib issued the plantation licence, including authorisation to clear the whole of the rainforest and to convert it into oil palm plantations without consulting the indigenous inhabitants in any way. In March 2011, indigenous plaintiffs went to court in Sarawak to sue for recognition of their native customary rights to use the land, which had previously been mapped with the assistance of the Bruno Manser Fund. The High Court of Sarawak, hearing the case in first instance, turned it down on formal grounds. The appeal by the indigenous plaintiffs is pending before the Malaysian Appellate Court at the time of writing.[31]

The release of the video by Global Witness a few weeks before the parliamentary election of 5 May 2013 led to a storm of indignation in Malaysia, and more than a million viewers watched the video on YouTube in a matter of weeks. Taib, however, managed to wriggle out of the public pressure. Neither the anti-corruption authorities nor the judiciary in Kuala Lumpur dare lay a finger on him.

TAIB'S CORRIDOR OF CORRUPTION

Quite apart from the clearing of tropical rainforest to make way for oil palm plantations, there is a new element in Taib's "development" master plan for Sarawak that is threatening to destroy Sarawak's tropical wilderness for the foreseeable future. Just as with the expansion of the oil palm plantations, here too corruption and oligarchy play a central role. The victims are once again the original inhabitants of Sarawak, in particular the Orang Ulu peoples living in the interior of the country—among them the Penan, Kayan, Kenyah, and Kelabit—, who are threatened with compulsory resettlement and even the complete eradication of their culture. Now that the majority of virgin forests have been chopped down, and plans are in full swing to replace large parts of the secondary forest with plantations, Taib plans to drown the river culture of the indigenous peoples by creating vast new reservoirs. The 2,400 megawatt Bakun Dam in Sarawak—which cost 2.4 billion US dollars and was commissioned in 2011—is already one of Asia's largest dams.[32] The area flooded by the Bakun reservoir is 696 square kilometres, slightly more than the area of Singapore.[33] The flattish topography of the area and the resulting vast reservoir has enormous social and environmental consequences. Ten thousand rainforest dwellers were forced to abandon their villages to make way for the mammoth dam and were transported to the bleak resettlement centre of Sungai Asap.

In the context of the Sarawak Corridor of Renewable Energy (SCORE), supported by the Malaysian federal government, Taib is planning a gigantic programme of industrialisation in Sarawak. Projected investment of 105 billion US dollars makes it the biggest energy project ever undertaken in Southeast Asia, and one of the most capital-intensive and ambitious government programmes anywhere in the world.[34] The underlying idea is simple. Enormous new reservoirs will collect the waters of

Borneo's powerful rivers and provide huge quantities of electricity, which will then be transported to Sarawak's coastal region across high-voltage power lines. There the cheap electricity will attract energy-intensive (and environmentally harmful) heavy industries, such as aluminium smelters, steelworks and glassworks.

The advantage of this plan, from Taib's point of view, is that his family stands to share in the pickings at every stage of SCORE: building the dams, constructing the high-voltage power lines, and developing the heavy industry. In contrast, the indigenous peoples of Sarawak lose out at every stage. They lose their land when it is flooded; they are forced to move to state-run resettlement centres; and they are inadequately compensated for their sacrifice.

One could be forgiven for thinking that this sort of "sledgehammer development policy" was a thing of the past, but Taib, a man of the 1960s, sensed huge new opportunities for making money. And he knows, from experience, that hardly anyone will dare stand in his way.

It was purely by chance that the plans for building the massive dams became public. Early in 2008, the organisers of an energy conference in Nanning, South China, mistakenly placed a confidential presentation by Sarawak Energy on their website, when, in fact, it was intended to be kept secret. Taib's brother-in-law Abdul Aziz Husain, in his then-capacity of director of the state-run electricity utility, had presented the Taib government's ambitious plans for completing twelve new hydroelectric power stations with a total capacity of 7,000 megawatts by 2020 (compared with a peak demand of less than 1,000 megawatts in 2008).[35] At that time, the Malaysian government was still planning to transport electricity more than 700 kilometres by an undersea cable from Bakun Dam, then under construction, to Peninsular Malaysia.[36]

Much has changed since them. Transporting electricity through the South China Sea to Peninsular Malaysia was deemed uneconomical; the Anglo-Australian mining giant, Rio Tinto Alcan, abandoned plans for huge aluminium smelting works in Sarawak; and, when the Bakun hydroelectric plant, which had cost billions to build, went on-line, it resulted in a vast oversupply of power in Sarawak.

The only thing that has not changed is Taib's outsized enthusiasm for dams. Even before the highly controversial Bakun Dam was completed, work began in 2008 on construction of Murum Dam in the upper reaches of the Rajang basin. With the whole site strictly cordoned off to keep out the public, two thousand Chinese and Pakistani workers under the direction of an Australian project manager set about work on the 141-metre-high dam that would generate an additional 900 megawatts of electricity.

The cement would be acquired from the Taibs, whose family business, Cahya Mata Sarawak, enjoys a cement monopoly. A social and environmental impact assessment of the 1.5 billion US dollar Murum project was not carried out until two thirds of the dam had already been built,[37] thus avoiding delays in the construction work. The result of such an assessment was a foregone conclusion anyway: The Murum Dam had to be built and the indigenous people living along the river had to move out. A "sweetener" was offered as compensation to the original inhabitants—compulsory resettlement in new houses, but the Taib family even managed to make money out of building them; two Taib family businesses, Naim Holdings and Cahya Mata Sarawak, were paid 220 million ringgits (around US$ 60 million) for constructing the Murum resettlement centre.[38]

Research carried out by the Bruno Manser Fund has shown that, if all the twelve dams planned until 2020 were to be built, an area of 1,600 square kilometres of tropical rainforest would disappear under water, and between 30,000 and 50,000 individuals belonging to 235 affected communities would be displaced.[39]

Taib's "Sarawak Corridor of Renewable Energy" is not only a serious environmental and social disaster—it is also a shamefaced misnomer. There are plans to use the same green cloak of renewable energy for five new coal-fired power stations with an annual consumption of 12.8 million tonnes.[40] Moreover, one of the priority sectors figuring within SCORE is to boost the output of crude oil.[41] It would be much more honest to abandon the term "Corridor of Renewable Energy" for this mammoth project and to speak instead of one huge "Corridor of Corruption".

NORWEGIAN FLOODS RAINFOREST

Taib's ambitious plans for hydroelectric and coal-fired power stations are being driven by a Norwegian citizen, Torstein Dale Sjøtveit, who has been at the head of Sarawak Energy since the end of 2009 as its CEO, a function in which he replaced Taib's brother-in-law as the undertaking's operational boss. Even under Sjøtveit's leadership, Sarawak Energy has remained firmly in the hands of the Taibs. Taib's cousin Hamed Sepawi is its chairman, and since the company is 100% owned by the Sarawak state, Taib in person—at least as long as he was chief minister—always had the last word.[42]

Sjøtveit has an annual salary of 1.2 million US dollars and is thus one of the best-paid managers in Malaysia. He likes to portray himself as an altruistic development aid worker—striving, at long last, to bring prosperity to the inhabitants of Sarawak.[43] In so saying, he glibly glosses over the fact that similar promises were uttered thirty years ago when the logging industry moved in, and that corruption and Taib's political stranglehold are today the principal obstacles to Sarawak's development.

In a blog published by Sarawak Energy, Sjøtveit displays tear-jerking pictures of his childhood in the Norwegian county of Telemark, and tells how his small home town of Rjukan managed to enter the age of industrialisation in 1911 thanks to the taming of a waterfall and the development of hydroelectric power.[44] Another photograph shows Sjøtveit affectionately shaking the hands of Penan children at the opening of a kindergarten.[45] Sjøtveit can, however, be a completely different man when the need arises, for instance when the subject turns to non-government organisations (NGOS), such as the Save Sarawak Rivers Network or the Bruno Manser Fund, which support Sarawak's indigenous inhabitants in their resistance to the Taib regime's plans for hydroelectricity. In writing to a Norwegian journalist in May 2013, Sjøtveit decried the supposed "lack of development" propagated by foreign NGOs in Sarawak as a "crime against the population at large and grossly undemocratic".[46] When, in September 2013, indigenous people set up a blockade to protest against their compulsory resettlement to make way for Murum Dam, Sjøtveit ascribed it to the "evil influence of foreign instigators".[47]

Sjøtveit rejects any responsibility for corruption in the dam business in Sarawak and maintains that throughout his life he has never issued any contract that had not been examined in a "transparent process applying the highest of standards".[48]

And yet, the facts contradict this. Within three years of Sjøtveit taking up his office as CEO, Sarawak Energy issued contracts worth more than 689 million Malaysian ringgits (roughly 200 million US dollars) to the Cahya Mata Sarawak, Naim Holdings, and Sarawak Cable companies—all of which have close ties with the Taib family—in connection with the dam projects.[49] In the aftermath of the Malaysian general election of May 2013 (see chapter 10), further contracts for power lines worth 618 million ringgits (185 million US dollars) were granted to Sarawak Cable, whose chairman is Taib's son Abu Bekir.[50]

When confronted with these allegations, Sjøtveit accuses his critics either of deception or of errors, which he fails to specify. "I have never been corrupt or involved in corrupt practices; not before I came to Sarawak, not while I have been in Sarawak, and I will not be in the future," Sjøtveit wrote to the Bruno Manser Fund in June 2013. As thanks for his service to the Taib regime, the governor of Sarawak bestowed the honorary title of "Datuk" (= "Sir") on Sjøtveit in September 2013.[51]

TAIB'S FOREIGN HELPERS

Sarawak Energy's CEO, Torstein Dale Sjøtveit, is just one of many foreign business people profiting from close ties with Taib. Other accomplices include lawyers, financial experts, accountants, and also managers with engineering qualifications, like Sjøtveit. At the end of 2011, in the context of its "Stop Timber Corruption" campaign, the Bruno Manser Fund published a blacklist in the format of a "Wanted" poster with the names of thirty individuals from nine countries, who were accused of providing financial, technical, or other services to Taib, and thereby of having supported or legitimised the despot's regime. Alongside the Norwegian Sjøtveit, this list of "Taib's Foreign Helpers" featured nine Australians, six Canadians, four British citizens and ten individuals of other nationalities.

The most prominent European on the list is Prince Albert II of Monaco. Through his public appearances with Taib, Albert II has given Taib international respectability. In April 2008, Albert II paid a state visit to Sarawak, during which he was accompanied by a private banker from Monaco and the Anglo-Greek property trader Achilleas Kallakis, who at the time was a member of the board of Prince Albert's environmental foundation.[52] That visit became a PR disaster for Albert II, not only because of Taib but because of Kallakis as well. In January 2013, a London court sentenced Kallakis to seven years' imprisonment for using falsified bank guarantees to gain possession of properties worth 750 million pounds (1.1 billion US dollars).[53] The visit to Sarawak had apparently been organised by Evelyne Genta, Monaco's ambassador to the United Kingdom (and wife of the late Geneva watch designer, Gérald Genta).[54] Ms Genta also played a central role in 2010 in setting up an "Islamic fashion show" sponsored by the Taib government in Monaco. Taib and Malaysia's prime minister, Najib Razak, were standing on the stage on 9 August 2010 when Albert II received a donation of 100,000 euros from the hands of Rosmah Mansor—the Malaysian prime minister's wife—for his environmental foundation, the *Fondation Prince Albert II de Monaco.*[55]

The most prominent American to make his way onto the list of Taib's helpers was the former FBI director, Robert Mueller. Mueller's name was included because of the FBI's decision to close their eyes to the corruption of the Malaysian despot when moving their Seattle head office into a building belonging to the Taib family (see chapter 1). Mueller was written to on this subject on numerous occasions but his office has never provided a reply.

One of the names from Australia on the list is that of James McWha, the former vice chancellor and president of the University of Adelaide. The university where Taib completed his law studies in 1961 not only granted its esteemed graduate and generous patron an honorary doctorate, but in 2008 went as far as to name a plaza on the university campus after him (see chapter 4). Professor McWha praised the despot for his "tireless work in helping to promote and strengthen the good relations between the two countries".[56]

For Taib and his clientele, personalities like Sjøtveit and McWha play a key role as gatekeepers between the criminal world of the kleptocrat and the global economic and political world. Taib's foreign helpers have been handsomely rewarded personally or with "charitable" donations. All he asks for in return is for them to avert their eyes.

LOSING SARAWAK'S RIVERS

The rapidly advancing transformation of the rainforest state of Sarawak into a "green wasteland" of oil palm plantations and reservoirs is not only threatening the destruction of Borneo's unique biodiversity, but is also robbing indigenous peoples of their rights to natural resources. The publicly accessible forest—the "capital of the poor"—is disappearing, giving way to plantations and energy projects under the control of a handful of timber barons and a tiny political elite.

For decades, the Penan and other communities have been protesting against the theft of their habitat by Taib's government, initially by blockading the logging roads, and, in the past decade, increasingly through courts, with lawyers belonging to the Sarawak Indigenous Lawyers' Association (SILA, see chapter 6). More than 200 cases have already been filed, and dozens have been concluded. The courts decide more often than not in favour of the indigenous communities.

The question, however, is whether Taib will respect these court judgments. One particularly sensitive issue is whether, along with the cultivated farmland, the much larger areas of communal forest *(pulau gala)* and longhouse communal lands *(pemakai menua)* will be recognised as indigenous territory too. At the end of 2012, one indigenous lawyer, Baru

Bian, sounded the alarm after a minister in the Taib government declared that these areas would not be recognised as indigenous land. Baru Bian accused the Sarawak government of contempt of court after two cases had made their way to the Malaysian Federal Court, where the court confirmed that the indigenous customary rights over communal forest and communal lands associated with longhouses had to be recognised.[57]

Another trick involves expropriating indigenous land for projects "in the public interest" and then handing it over to Taib's allies. Dam projects are particularly suitable for this, since it is possible to take not only the land needed for the reservoir itself but also any land within the project perimeter. This happened within the catchment area of Bakun Dam, where logging and plantation corporations, such as Shin Yang and Rimbunan Hijau, were permitted to clear huge areas of forest and establish plantations. Despite promises by the Malaysian government that the forest was to be preserved in the catchment area, these two big companies were given licences for more than 300,000 hectares of plantations in secretive agreements.[58] Lawyer Baru Bian is afraid that the Taib government's latest dam projects may also serve as a pretext for expropriating indigenous land.[59]

Considering the negative experience of the Bakun Dam resettlement, which led to abject poverty, the indigenous people are fully aware of the possible consequences of Taib's latest projects for dams. In particular, in the densely populated Baram region—the traditional territory of the indigenous peoples of the Kenyah, Kayan and Penan—a resistance movement has been gaining pace since the end of 2011 and is supported by international organisations such as the Bruno Manser Fund, International Rivers, the Borneo Project and the Rainforest Foundation Norway.

The Save Sarawak Rivers Network (SAVE Rivers) and the Baram Protection Action Committee are channelling their commitment through information events, public statements and protest actions against the destructive dam projects. "Reservoirs and hydroelectric power stations lead to poverty, not to development," the indefatigable campaigners write on their website. "Sarawak's dams have already driven 12,000 people out of their homes. In the resettlement centres for the Bakun and Batang Ai dams, many indigenous people are living in poverty, without work or sufficient arable land, and with poor access to education and healthcare."[60]

The independent anti-dam movement is presided over by Peter Kallang, a retired engineer, who previously worked for Shell and is also a committed Catholic. The level-headed but determined Kallang gives a straight answer when asked who would benefit from the planned dams: "In Sarawak it is only our rulers who are profiting from these dams. For us indigenous people, they are a threat not only to our livelihood but to our cultural heritage as well."[61]

In the lead-up to an international hydropower congress in Kuching in May 2013, which Sarawak Energy had devised as a propaganda event for Taib's dam programme, Peter Kallang and the Bruno Manser Fund wrote critical letters to the International Hydropower Association (IHA). The hydropower lobby organisation and Sarawak Energy responded immediately by excluding the indigenous representative Kallang and Annina Aeberli (an employee of the Bruno Manser Fund) from the congress, although both had paid the steep congress fee of US$ 1750. Kallang filed a criminal-law complaint against the IHA and Sarawak Energy at the nearest police station.[62] Under international pressure, Kallang and Aeberli were then admitted to the subsequent sessions of the congress, at which they were able to speak and make critical comments, despite the resistance of the representatives of the Taib government.[63]

While delegates from sixty countries were debating in the air-conditioned congress centre the progress made by the "sustainable" use of hydropower, 300 indigenous protesters against Taib's dam plans had gathered outside with clenched fists and banners saying "Stop Baram Dam", "Bakun is enough", and "Respect native rights". They addressed a demand to the IHA director, Richard Taylor, that the hydropower lobby group distance itself from Sarawak Energy's plans for dams and that it eject Taib's Norwegian confidant Torstein Dale Sjøtveit from its board.[64] The dam opponents were particularly enraged that the IHA was holding its gathering in a congress centre belonging to the Taib family (the Borneo Convention Centre Kuching), which had both Taib's son Abu Bekir and sister Raziah Geneid sitting on its board of directors.[65]

Negative press and quizzical participants were certainly not the way that Taib had imagined his hydropower congress in Kuching would begin. At the end of the conference, Agence France Presse circulated a report with the unambiguous title of "Outrage grows over scandal-tainted Malaysian leader", which was taken up by the media throughout Southeast Asia. It is hard to imagine a more clearly-worded agency communiqué.[66]

There is still much work to be done by the land rights lawyers in Sarawak and the campaign managers of SAVE Rivers (including, in particular, Mark Bujang and Philip Jau, along with Peter Kallang). The positive public response to their work provides hope that it may still be possible to halt Taib's destructive plans and stop the almost total transformation of the once "Fair Land Sarawak" into a green wasteland of plantations and reservoirs.

10

RAINFORESTS WITHOUT CORRUPTION

After thirty-three years in power, Taib resigned as chief minister in early 2014 to become governor of Sarawak. But is this a sign of real change? Considering the scale of corruption and environmental destruction in Sarawak, the international community must recognise the crimes that have been committed by the Taib family. There is another sign of hope, which is being carefully nurtured by the Penan with their autonomously administered "Penan Peace Park".

SARAWAK AT THE CROSSROADS

I write these lines four years after Clare Rewcastle and I travelled to California to hear the insider's account of Taib's real estate empire. Much has happened since we collected the stacks of documents. The Bruno Manser Fund has launched its campaigns against corruption in the timber trade in Malaysia ("Stop Timber Corruption") and against Taib's plans to build dams in Borneo ("Stop Corruption Dams"), and both of these had marked international resonance. To coincide with the 30th jubilee of Taib's assuming office as chief minister, the Bruno Manser Fund coordinated worldwide protests in the spring of 2011 in Ottawa, Seattle, San Francisco, London, Berne, Sydney, Adelaide, and Huonville (Tasmania).[1] In December 2011, the Canadian television station *Global Television* reported in detail for the first time ever about the money laundering accusations against the Taib circle.[2] A year later, a report by the Bruno Manser Fund analysing the business transactions of the individual family members triggered a strong reaction in Malaysia.[3]

Clare Rewcastle has set up the investigative *Sarawak Report* blog and has founded the short-wave radio station *Radio Free Sarawak*, which provides the native population in the Sarawak interior with independent news.[4] This commitment earned the prestigious "Free Media Pioneer Award" of the International Press Institute in Vienna for Rewcastle and her clandestinely-operating team of presenters, Peter John Jaban ("Papa Orang Utan") and Christina Suntai, in May 2013.[5] The commitment to free press is especially important, since all the newspapers and the electronic media, with the sole exception of the Internet, are subject to strict government censorship in Malaysia.[6]

There has also been movement inside Malaysia itself. A newly created Facebook group, "Taib must go", has been joined by more than 30,000 individuals, and corruption is now being talked about more openly and frequently in Sarawak than ever before—a development unthinkable only a few years ago. Revelations by various whistle-blowers and business partners of the Taib circle, through Clare Rewcastle's *Sarawak Report*, have pushed the ugly truths into the spotlight. In June 2011, the Malaysian anti-corruption commission (MACC) announced that it had initiated an investigation into Taib.[7] Taib's reaction was not slow in coming. He placed Clare Rewcastle on a blacklist of undesirable individuals. When she flew to Kuching in July 2013, she was refused entry to Sarawak and forced to take the next flight back to Singapore.[8]

The most significant development, however, came from the Malaysian government. After thirty-three years in power, Taib resigned as chief minister of Sarawak at the end of February 2014 to become the state's governor, as his uncle had decades earlier. Government insiders insist that Taib was forced to step down by the federal government in Kuala Lumpur, and that the ruling Barisan Nasional coalition didn't allow his son, Abu Bekir, to run in a by-election to replace him. The new chief minister, 70-year-old Adenan Satem, a former federal deputy minister, is widely looked upon as an interim figure. As governor, Taib's new role is supposed to be purely ceremonial. But it remains to be seen whether he has truly ceded power or whether he continues to pull the strings from behind the scenes.

Taib's removal from the office of chief minister followed his party's mediocre results in the last elections at regional level in 2011 and at national level in 2013. While the despot's mandate was secured by an easily controllable rural electorate, he lost the politically important urban middle class to the opposition. As many times before, Taib bought rural votes on a massive scale, and it is exceedingly likely that fraud occurred during vote counting, in particular for the close seats in the Baram region, where emotions were running high on account of the projected dam. It is, nevertheless, a paradox that the Taib regime owed its power over many years precisely to those who stood to lose most from its policies. The dependence of the impoverished indigenous peoples on the government is far too great for many of them to dare rebel against Taib. "Never go against the *towkay*" is how Taib's loyal lackey, James Masing, formulated it,[9] and that would appear to be a credo that many Sarawak natives have very much taken to heart.

Since the end of British colonial power in 1963, Taib has been a minister for fifty-one years and chief minister of Sarawak for thirty-three. His rule has been extraordinarily stable—and extremely corrupt. Nonetheless, there are signs that times are about to change in Sarawak, if for no other reason than because the 78-year-old Taib won't live forever.

Taib's involuntary departure from the office of chief minister is a strong indication that the federal government in Kuala Lumpur is no longer willing to allow one family to reap the lion's share of the state's economic resources. Prime Minister Najib Razak must also have felt increasingly uneasy that his government depended more and more on Taib's votes than ever before. Without Sarawak's twenty-five parliamentary seats (out of 133 for the ruling *Barisan Nasional* coalition) in the May 2013 general election, Najib would have lost office to his long-time rival, Anwar Ibrahim. For the first time in the country's fifty-year history, a majority of Malaysians voted in favour of the political opposition. The close election result made Sarawak the political battleground where the country's political future will be decided. Further conflict between federal and state government actors is inevitable.

Taib built his system of rule essentially on networks of dependency by means of his strict control of resources and information. That significant parts of the Sarawak interior still have no decent roads and substandard education and healthcare facilities is in all likelihood the outcome of a deliberate strategy. The better educated and the more economically independent the rural population is, the less it will be reliant on the ruler's goodwill. Certain road projects have been packaged as electoral promises by the Taib government on many occasions in the past, but for many years there has been no sign of their fulfilment.

There are two developments that are slowly but surely eroding the power base of Taib circle. Firstly, the natural resources that are easiest to exploit—the most valuable tropical timbers and the best land for oil palm plantations—are virtually exhausted. This makes it more difficult for the Taib regime to glean profits and pass on a slice to its political allies. This leads to dissatisfaction. In many ways the over-ambitious dam projects are a desperate attempt to find lucrative new sources of kickbacks. At the same time, however, these projects are giving rise to new conflicts, as is shown in the vehement dispute over the Baram Dam.

The second, and probably even more significant, threat to the Taib circle lies in the digital revolution, and the increasing powerlessness of rulers to control the flow of information. Every new mobile telephone and every additional Internet connection in the interior of the country threatens Taib's control over his power base, the impoverished rural population. The better informed the people are, the less readily they can be manipulated. In addition, rapid globalisation of information is making it more difficult for Taib to shield his feudal state from scrutiny by the rest of the world.

In a speech delivered in May 2013 to the state assembly, the Sarawak regional parliament (which is more like a sultan's court than a genuine parliament), Taib made an issue out of this threat and complained about the foreign criticism of his handling of the affairs of government: "Neither the State Government nor I am accountable to the foreign NGOs and foreign reporters or broadcasters. Neither would I allow them to try to hold me accountable for matters concerning my administration of the State. To respond or react to what they wrote about me would be to acknowledge that they have a right to interfere in or participate in the affairs of this country." Taib went on to accuse the foreign NGOs of wanting to re-colonise Sarawak. In doing so, Taib attacked, by name, the Bruno Manser Fund, Clare Rewcastle's *Sarawak Report, Radio Free Sarawak*, and Global Witness, which, he claimed, were specialised in "making malicious accusations" in order to "drag the Sarawak government and Taib through the mud". He continued: "Their motivation is purportedly their concern about the governance of the State and the alleged plight of the indigenous people but if what they do is subjected to impartial and careful analysis, their hidden agenda could clearly be discerned and that is to cause political instability and halt the development momentum and the economic growth that Sarawak has achieved since Malaysia Day, under its own democratically elected Government. (...) To me, this is a disguised attempt to introduce a disgusting form of a recolonialization [sic]."[10] If there had been need of evidence that the increasing criticism of his corrupt rule was taking a toll on the thin-skinned Taib, there it is!

WHY SARAWAK HAS TO MATTER TO US ALL

But why bother about Sarawak at all? Isn't this a Malaysian problem, to be dealt with by Malaysians? Shouldn't we simply avert our eyes and follow the examples set, for instance, by the FBI or the University of Adelaide? There are good reasons for not doing so.

• **FIRSTLY,** the Taib regime in Sarawak has continuously and systematically violated the human rights of the indigenous peoples of Sarawak, as laid out, for instance, in the United Nations Declaration on the Rights of Indigenous Peoples (UNDRIP).[11] Fundamental rights, such as the right to own land, the right to self-determination, and the right to have their culture respected are violated every single day. Although UNDRIP lays down that the use of indigenous land for public purposes requires the free, prior, and informed consent ("FPIC") of the communities affected, the Sarawak rulers often decide on the fate of vast expanses of indigenous land without any form of consultation or transparency.

What is particularly objectionable is that, more than fifty years after Malaysia gained its independence, many indigenous people still do not have valid identity documents, and around one third of the citizens of Sarawak do not have their names on the electoral register, and are thus deprived of their fundamental political rights.[12] Malaysia is one of only a few countries in the world that have not ratified the International Covenant on Civil and Political Rights (in which it finds itself in the company of countries like Saudi Arabia, Myanmar, and North Korea), a key human rights treaty protecting individual liberties.[13] If a country violates human rights instead of protecting them, it behoves the international community to exercise meticulous scrutiny and to ensure that those in power are called to account.

• **SECONDLY,** corruption—and the money laundering that goes hand in hand with it—is a criminal offence in nearly every country on earth and is amongst the few crimes that can be prosecuted internationally. What that means is that the citizens or corporations of one country can be made to answer for their corrupt actions in another country too. This was the case, for instance, at the end of 2011, when the Swiss branch of the French Alstom group had to pay a fine of 2.5 million Swiss francs (US$ 2.7 million) and forfeit 36.4 million Swiss francs (US$ 40 million) of profit to the judicial authorities because the group had bribed public officials in Malaysia, Latvia, and Tunisia.[14] In August 2013, the Swiss Attorney General opened a criminal investigation into UBS on suspicion that suspected corruption proceeds from the tropical timber trade in Sabah, Malaysia, had been laundered by its branches in Singapore and Hong Kong (see chapter 7).[15]

Malaysia (along with 166 other countries) has ratified the United Nations Convention against Corruption (UNCAC), and has thereby assumed a commitment under international law to combat it. In the preamble to the convention, which was adopted by the UN General Assembly in 2003, corruption is defined as a threat posed to "the stability and security of society undermining [...] democracy, ethical values and justice and jeopardising sustainable development and the rule of law."[16] Various convention articles refer specifically, for instance, to embezzlement by public officials, trading in influence, abuse of functions, and illicit enrichment as fields in which there is a particular need for action.[17]

Ironically, a clear description of the dangers of corruption is to be found on the website of the FBI office in Seattle—located in the Taib building—"Corruption in government threatens our country's democracy and national security, impacting everything from how well our borders are secured and our neighborhoods protected to verdicts handed down in courts and the quality of our roads and schools. It wastes billions of tax dollars every year."[18] In the light of the experience in Sarawak, it could certainly be added that corruption constitutes a massive threat to the environment too. Politicians who fund their electoral campaigns with bribes from logging corporations are never going to defend the rainforests, not even if the last very tree has to go.

The dirty money from corruption in Sarawak, running into billions of dollars, is having global effects. Taib's illegal profits have mostly flowed out of Malaysia—and continue to do so. It must be strongly suspected that this flow of money is connected with illegal actions—and market distortion—in many countries, and it is likely that they are handled to a significant extent by financial service providers, law offices, and construction companies with transnational activities, existing in a grey zone of dubious legality. These global players have an unfair advantage in comparison to businesses whose whole operation is entirely above board and who refuse to work with corrupt despots. The Taib family's investments in real estate worth hundreds of millions of dollars in Canada, the USA, the United Kingdom, and Australia has had direct impacts on the integrity of those countries' property markets. In the case of the American real estate manager Ross Boyert (see chapter 1), there are grounds for suspecting that the Taib family arranged and paid for a stalking campaign against their former employee, driving him to suicide. Another incident reported from Australia is that a former business partner of the Taib family, the building contractor Farok Majeed, was threatened and forced to live in hiding.[19] Taib's brother Onn is accused of having cheated the Australian tax authorities out of several million US dollars.[20]

The export of illegal assets from Malaysia is not limited to the Taib family. According to calculations made by the US organisation Global Financial Integrity, Malaysia has one of the world's highest rates of capital flight. In 2010 alone, Malaysia exported 64 billion US dollars of illicit assets. Taking the whole decade from 2001 to 2010, the total is estimated to have been 285 billion US dollars.[21] Malaysia thus occupied the third position on this shameful global league table, after China (2.7 trillion US dollars) and Mexico (476 billion US dollars). The transfer of illegal assets on this scale is bound to have a major impact on the integrity of the international financial system and its players.

• **THIRDLY,** Taib and his regime bear principal responsibility for the destruction of Borneo's rainforest, an ancient and unique biosphere with a vast number of native species of fauna and flora, and an indigenous culture that has adapted to it. It is not by chance that it was precisely in Sarawak in 1855 that the naturalist and explorer Alfred Russel Wallace gained fundamental insights into the evolution of the species in parallel to Charles Darwin (see chapter 2). Borneo's biodiversity belongs among the greatest natural treasures on Earth. Should the destiny of these animal and plant species and a millennia-old indigenous rainforest culture really be left to the whim of a despot?

The Sarawak rainforest is far too important a part of the world's natural heritage for it to be left in the hands of the Taib circle, which has already wrought irreversible damage in the span of a single generation. Major parts of the peat swamp forests in lowland Sarawak, for example, have already been destroyed. They form the natural habitat of proboscis monkeys, orang-utans, and the Sarawak surili (one of the world's rarest primates).[22] Peat swamp forests store particularly large amounts of carbon, and destroying them releases vast amounts of greenhouse gases into the atmosphere. Deforestation is advancing very rapidly in Sarawak. Within just one five-year period, from 2005 to 2010, a third of Sarawak's peat swamp forests was cut down. In what used to be the rainforest state of Sarawak, it is virtually impossible to find any pristine virgin forest today, fifty years after it joined Malaysia.[23]

Politically, Taib was always primarily a power monger and advocate for his family. Despite numerous promises, he was never seriously ready to make a move towards sustainable development and reconciliation of environmental, social, and economic concerns.

Of course, in managing Sarawak's forests, it is possible for Taib to invoke the principle of national sovereignty over the natural resources as one of the central tenets of international law. Furthermore, the Malaysian constitution of 1963 lays down explicitly that Sarawak and Sabah, the two East Malaysian states on Borneo, enjoy complete autonomy in questions of the exploitation of their resources. That cannot, however, be a blank cheque for the wilful destruction of the forest, and the distribution of state land to members of his family and political followers.

One important principle in international environmental law is that of "common but differentiated responsibility" for the environment. This holds that we all share responsibility for the global environment, but that we each have a responsibility to be active in that place where we hold particular responsibilities and powers. It is a principle that developing and newly industrialised countries like to invoke in connection with the question of who is to pay for cleaning up environmental damage (the finger points at the industrialised countries, because they have money and have been contaminating the environment for longer). The same principle, however, applies also in the opposite direction. Those countries whose biodiversity is particularly rich and of global significance ought to be especially active in protecting it.

This imposes a clear obligation on Malaysia and other countries with tropical rainforests. Malaysia, in particular, which has very considerable financial resources available to it as an oil-producing country, must be put under much greater pressure by the international community to face up to its responsibility.

By signing in 2007 the "Heart of Borneo" declaration initiated by the WWF, the governments of Malaysia, Brunei, and Indonesia recognised the particular value for humanity of Borneo's rainforests.[24] When it comes to putting this into practice, however, the Malaysian federal government in Kuala Lumpur has allowed its political ally Taib to do as he wishes in attacking the rainforest, while impeding action by the country's criminal justice system.

FAILURE OF THE INTERNATIONAL COMMUNITY

The Taib case illustrates clearly the shortcomings in the Malaysian and international legal systems. Given his political power in Sarawak and his importance for the federal government in Kuala Lumpur, Taib has long enjoyed an "unimpeachable" status in Malaysia, under which he has been able to ride rough shod over national and international law. He has stubbornly ignored rebukes from institutions such as the Malaysian human rights commission, SUHAKAM, for his disregard of indigenous land rights. He has publicly referred to the Malaysian anti-corruption commission, which is carrying out an investigation into him, as "cheeky and dishonest" and announced his refusal to cooperate.[25]

As long as the ruling *Barisan Nasional* coalition remains in power in Sarawak and Malaysia (which it has monopolised ever since independence) and as long as the judiciary toes the political line, Taib has little to fear in his own country. The question that must be asked is whether the international community and the international legal system are able and willing to call to account a blatantly criminal ruler from a newly industrialised country.

In the context of its "Stop Timber Corruption" campaign, the Bruno Manser Fund wrote to the governments of Canada, Australia, the United Kingdom, Jersey, Germany, the USA, Switzerland, and others in 2011, providing them with concrete evidence of business or personal ties between the Taib family and the countries concerned. It called on the heads of government or the responsible ministers to instigate investigations and to freeze any assets the Taibs had in their territories. With the exception of the USA, each government promised to forward the information to the competent authorities. Generally speaking, however, the replies were sobering.

The then Canadian Minister of Finance, James Flaherty, expressed his thanks for the information and stressed that Canada was actively involved in international initiatives to combat corruption and that it was participating in the anti-corruption plan of the G20 countries.[26] In a separate letter, the Royal Canadian Mounted Police expressed its respect for the Bruno Manser Fund on account of its work to protect the tropical rainforest and the indigenous peoples. It regretted, however, that it was not able to express an opinion on the Taib case.[27]

The Australian Department of Foreign Affairs and the Australian Federal Police emphasised that they took the implementation of the UN convention against corruption very seriously, but that there was insufficient evidence in the Taib case for them to act. They requested the Bruno Manser Fund provide hard evidence that Taib real estate in Australia had indeed been financed out of the proceeds of corruption.[28]

The responsible minister at the British Foreign and Commonwealth Office, Jeremy Browne, wrote to the BMF saying that they were going to keep a watch on the investigation by the Malaysian anti-corruption commission. The United Kingdom had just signed a memorandum with Malaysia on combating international crime, which included close co-operation in the field of money laundering.[29]

The financial supervisory authority in Jersey—a British Crown dependency in the Channel Islands—wrote back that it would be better for the BMF to address its complaint against Deutsche Bank (on suspicion of having laundered money for the Taib company of Sogo Holdings Limited) directly to Deutsche Bank in Jersey itself.[30]

At first sight, the most encouraging reply was the one the BMF received from Germany. The federal ministry of finance launched an enquiry into the business relationship between Deutsche Bank and the Taib family to determine if it was in compliance with German regulations against money laundering.[31] However, BaFin, the Federal Financial Supervisory Authority entrusted with this enquiry, also concluded that there was no basis for it to intervene.[32]

The situation is clear: No national government is willing to risk its relations with a leading figure in the government of another sovereign state because of corruption. The threshold needed for interference in the affairs of other countries is very high, unless important interests of one's own happen to be at stake. As long as the Malaysian federal government and the country's judiciary let Taib continue with impertinence, it is unlikely that any government will be stirred into any sort of action.

If the national governments are not going to act, little scope is left for action by those multilateral organisations whose remit it is to combat corruption and environmental crime. That is the clear message contained in a letter from Interpol (the world's largest international police organisation), to which the Bruno Manser Fund sent a copy of its letter to the judicial and police authorities in Malaysia, calling on them to place Taib and thirteen members of his family on its list of wanted persons and to have them arrested. Interpol, which is based in Lyon, France, is running an active programme to combat illegal logging and other environmental crimes.[33] Interpol wrote back: "Please be (...) advised that Interpol's assistance may be activated only by the decision of the competent domestic authorities and at the request of the national contact point—national central bureau (NCB). Should you wish to trigger criminal proceedings against the individuals at stake, you are kindly advised to contact the national police and/or judiciary."[34]

Interpol's advice to the Bruno Manser Fund was in vain, as that is precisely what the BMF had already tried to do in filing a detailed criminal-law complaint by registered mail, enclosing numerous items of evidence, and demanding that the Malaysian authorities arrest the Taib circle and instigate criminal proceedings against its members.[35] However, not a single representative of Malaysia's federal police, its anti-corruption commission, or the office of its federal attorney general responded to the letter from the BMF.

As long as neither Malaysia nor any other state is willing to act, Interpol and other multilateral organisations have their hands tied.

MORE RECENT INITIATIVES AGAINST ILLEGAL LOGGING

It is not only in Sarawak that corruption and illegal logging are a massive problem. The World Bank puts a figure of 10 to 15 billion US dollars on the annual volume of trade in illegally logged timber.[36] David Higgins, the head of the Interpol department for combating environmental crime, goes even further than that; he estimates the value of timber illegally logged every year at more than 30 billion US dollars.[37] It would be naïve to believe that such huge amounts of timber could be felled and smuggled abroad without the knowledge and acquiescence of authorities in the countries of origin. It is more likely that, as with Taib in Sarawak, there are many other countries in which leading politicians and senior officers in the forestry authorities make money out of the clearance of the tropical forests. Examples include Papua New Guinea, Indonesia, and the Solomon Islands.[38]

When Interpol, in cooperation with the United Nations Environmental Program (UNEP), launched a new initiative to combat illegal logging in 2012, the international police organisation justified it in the following words: "The criminals responsible for illegal logging are destroying bio-diversity, threatening the livelihood of those reliant on forest resources, and contributing directly to climate change. With corruption, violence, and even murder tied to illegal logging, this type of crime can also affect a country's stability and security."[39]

Much of this reasoning applies not only to illegal logging but also to the perfectly legal, economically motivated, and apparently blameless ex-ploitation of timber resources. A major contributory factor in the world-

14 LARGE MALAYSIAN COMPANIES WITH TAIB FAMILY OWNERSHIP (2011)

Name of company	Net assets (Ringgit)	Taib family stake	Taib family net assets (Ringgit)
Cahya Mata Sarawak (CMS) Berhad	2,451,501,000	56.8%	1,392,452,568
Achi Jaya Holdings Sdn. Bhd.	550,075,412	100%	550,075,412
Ta Ann Holdings Berhad	1,397,121,272	35.30%	495,978,051
Custodev Sdn. Bhd.	1,578,782,271	25%	394,695,567
Lembah Rakyat Sdn. Bhd.	286,454,208	99.50%	285,021,937
Perkapalan Damai Timur Sdn. Bhd.	387,017,150	60%	232,210,290
Naim Holdings Berhad	1,076,687,000	16%	172,269,920
Sarawak Plantation Berhad	495,446,000	30.45%	150,863,307
Sanyan Holdings Sdn. Bhd.	109,567,757	86.25%	94,502,190
Titanium Construction Sdn. Bhd.	94,142,694	60%	56,485,616
KBE (Malaysia) Sdn. Bhd.	91,153,662	60%	54,692,197
Sarawak Cable Berhad	142,950,414	32%	45,744,132
SIG Gases Berhad	93,612,796	18%	16,850,303
Smartag Solutions Berhad	34,447,350	30.60%	10,540,889
Total Ringgits			3,952,382,379
Total US Dollars			1,253,928,419

Source: Companies Commission of Malaysia

wide plundering of the tropical rainforest, however, is that the dividing lines between legal and illegal logging are particularly fuzzy. If the criteria for issuing and exploiting logging concessions and plantation licences occur without transparency and are subject to the payment of massive bribes—as in Sarawak—then it seems barely appropriate to speak of legal logging at all.

That is also the view taken by the United Nations Office on Drugs and Crime (UNODC). In analysing the threat posed by transnational organised crime in East Asia and the Pacific, the UN agency arrived at the conclusion that nearly all the illegally logged timber is assimilated into the lawful trade (and thus "laundered") while still in its country of origin. In that way, illegal timber can be exported officially. Illegal logging, moreover, is "mostly carried out by companies with a public face, often with international shareholders, who are already involved in the licit trade. Logging becomes illegal when the permits are acquired through bribery, or where protected species are involved, or where the harvesting takes place outside the agreed concession."[40]

On the basis of this definition, UNODC estimates that about half the total volume of timber exported from Sarawak is illegal. It also records its estimate that the highest percentages of illegal timber exported by the East Asian/Pacific region are by Papua New Guinea (90%) and the Solomon Islands, Cambodia, and Myanmar (each at 85%). These are all countries in which the timber corporations from Sarawak are or have been active (see chapter 8).[41]

Tellingly, despite many years of drawn-out efforts and negotiations, Malaysia has failed to qualify for a voluntary partnership agreement (VPA) with the European Union's Action Plan on Forest Law Enforcement, Governance and Trade (FLEGT), which was designed to prevent the import of illegally logged timber. The chief grounds are the dubious circumstances in Sarawak, and the Malaysian government's refusal to allow impartial consultations with non-government organisations, and associations of indigenous peoples.[42]

The EU's FLEGT action plan is one of the most hopeful signs in the struggle against illegal logging. Through bilateral negotiations and

TAIB COMPANIES WORLDWIDE

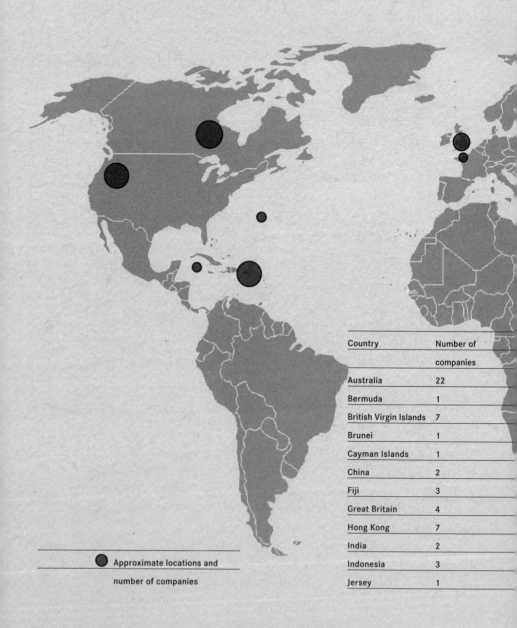

Country	Number of companies
Australia	22
Bermuda	1
British Virgin Islands	7
Brunei	1
Cayman Islands	1
China	2
Fiji	3
Great Britain	4
Hong Kong	7
India	2
Indonesia	3
Jersey	1

● Approximate locations and number of companies

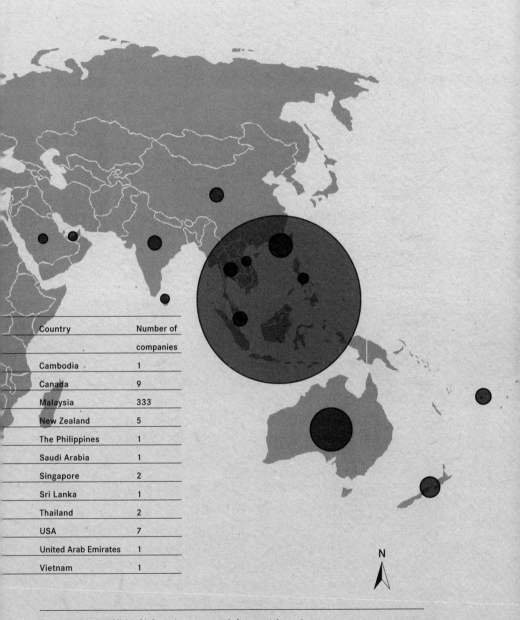

Country	Number of companies
Cambodia	1
Canada	9
Malaysia	333
New Zealand	5
The Philippines	1
Saudi Arabia	1
Singapore	2
Sri Lanka	1
Thailand	2
USA	7
United Arab Emirates	1
Vietnam	1

N

Source: BMF 2011 (Additional information: www.stop-timber-corruption.org)

long-term technical support, the European Union is helping the countries producing tropical timber to reform their timber sector and ensure the legality of the timber exported. In exchange, the participant countries are granted privileged access to the EU timber market.[43] A new European timber regulation, which came into force in spring 2013, defines new due diligence duties for the timber trade and is intended to make it more difficult to import illegal timber.[44]

In recent years, other major economies, such as the USA and Australia, have also enacted new laws intended to prevent the import of illegal timber and to make it a punishable offence. However, it remains to be seen to what extent the broadening of the US Lacey Act (2008) and the Australian Illegal Logging Prohibition Act (2012)[45] are really going to be effective instruments against the illegal timber trade. Two of the biggest difficulties are in proving that the timber is of illegal origin—since the relevant documents are generally counterfeits—and in making the difficult distinction between legally and illegally logged timber in international trade.

Will these measures against illegal logging suffice to protect the rainforests when the global demand for tropical timber remains at a high level? Recent ecological research casts fundamental doubt on whether the timber industry in the tropics is capable of maintaining the forests, given that harvest cycles of 30–40 years are far too short, and merely constructing forest roads often causes long-term damage to forest ecosystems. One Australian research team has forecast that the production of tropical hardwoods is soon going to show declining returns on account of over-exploitation after going through a production maximum for tropical timber (in other words: a similar pattern to that detected for crude oil).[46]

There are not too many substantive grounds for optimism, according to a report published in 2011 by the International Tropical Timber Organisation, which has taken on the task of encouraging sustainable forestry in the tropics. In 1990, the ITTO, which was set up under the patronage of the United Nations, set itself an ambitious "Year 2000 Objective" that all tropical timber exports should come solely from sustainable forestry by that year.[47] Twenty years later, the ITTO was forced to admit that still less than 10% of the 403 million hectares of tropical forest exploited for forestry around the world were being managed in a sustainable way.[48]

THE PENAN'S LAST HOPE

Is there any hope at all left for Borneo's rainforest? The lesson to be learnt from the history of Sarawak under Taib's rule would seem to be that "development" at any price is bound to go hand in hand with the destruction of the forests and an end to indigenous cultures. That does not necessarily have to be the case, however, especially since "development" as propagated by Taib benefits only a tiny elite and could not by any figment of the imagination be described as sustainable. Moreover, Sarawak is an extreme example of how one man and his family are snatching the natural resources of a whole country, and appropriating them for their personal enrichment.

When Bruno Manser left Sarawak to return to Switzerland in 1990 after living for six years with the Penan, the logging front in the rainforests there was advancing faster than ever. Manser resolved to dedicate his life to fighting on behalf of the Penan. He created the Bruno Manser Fund to provide support for his friends in the rainforest.

At the first annual general meeting of the new association, Manser described how urgent the situation in Borneo had become, given the rapid incursions of the logging companies: "The situation in Sarawak has deteriorated. If things keep going on like this, there are going to be no primary forests left in 6–7 years' time; twelve square kilometres are falling victim to the chainsaws every single day," Manser warned. "The situation is hopeless, unless something is done outside of Sarawak."[49]

More than twenty years later, the situation in Sarawak has deteriorated still further—radically so. The Taib government's change of course away from the unsustainable timber business to the palm oil economy has meant that between the 1990s and the present, more than a million hectares of rainforest have been converted to plantations. The government's plans envisage doubling the area of land under oil palm cultivation to two million hectares by 2020. In addition, so called "planted forests" are being created on a further six million hectares, which is another name for industrial timber monocultures.[50] If these plans come to fruition, then in the near future two thirds of Sarawak will be covered in monocultures and the former rainforest country will have been degraded to a plantation state.

The fact that pristine forest can still be found today outside the meagre protected zones—despite the Taib government's plans to remove it—is thanks to the courageous resistance by indigenous Penan communities. Since the 1990s, they have continuously and determinedly resisted the compulsory "development" prescribed by the Taib government, and have prevented the bulldozers from getting through. In particular, the well-organised Penan communities in the Upper Baram region, who have been settled for longer, have managed to save nearly 100,000 hectares of rainforest from the chainsaws, thanks to systematic blockades of logging roads.

That might appear to be no more than a modest success, but at least for those directly affected it means that they can continue to live on their ancestral lands, to practise farming and to go hunting.

"The resistance has been worthwhile" the late Penan headman and land rights plaintiff, Kelesau Na'an, told me nearly ten years ago. His village, Long Kerong, in the traditional East Penan territory on the Selungo River is today still a wonderful oasis in the virgin forest.

Long Kerong is one of eighteen Penan communities which joined forces at the end of 2009 and proclaimed the "Penan Peace Park", a self-governing nature reserve covering 163,000 hectares of rainforest in the Upper Baram. Sustainable farming, tourism, and protection of the forest are to make it possible for the Penan to live their lives there in self-determination—and for their culture to survive. The central element in this project is that the Penan want to take their future into their own hands.

The Penan's successful resistance to deforestation needs to continue if future generations are also to know the natural treasures of Borneo's rainforest. If that is to happen, the indigenous peoples of Sarawak need the determined support of the international community—not only from NGOs, but also from governments and responsible economic players.

The fundamental change must, however, come from Malaysia itself and must take place at the political level. Taib's removal as chief minister is a first, important step in the right direction. It is up to his successors to correct the state's course of action and the government's condescending attitude towards its indigenous peoples. Now, the Malaysian judiciary and anti-corruption authorities need to live up to their responsibility. While it is good that Sarawak's "Last Rajah" has finally stepped down, he does not belong in the governor's residence—he belongs in jail.

ENDNOTES

01 FOLLOW THE MONEY

1 An analysis of the Taib family's assets is to be found in Bruno Manser Fund, *The Taib Timber Mafia: Facts and Figures on Politically Exposed Persons (PEPs) from Sarawak, Malaysia* (Basel, 2012).

2 Jane E. Bryan et al., "Extreme Differences in Forest Degradation in Borneo: Comparing Practices in Sarawak, Sabah, and Brunei," *PLoS ONE* 8, no. 7 (2013), doi:10.1371/journal.pone.0069679.

3 Clare Rewcastle has published a large part of the documents handed over to us by Ross Boyert and also a video interview with the whistle-blower on her *Sarawak Report* blog, http://www. sarawakreport.org/. Some of the documents have also been included on the Stop Timber Corruption campaign pages of the Bruno Manser Fund, http://www.stop-timber-corruption.org/resources/.

4 Before Ross Boyert joined Sakti, one Taib building in San Francisco was forced into auction and a 3 million US dollar investment had to be written off. See letter of November 20, 2006, from Ross Boyert to Taib, http://www. sarawakreport. org/2010/10/ taib-handed-rockefeller-100-million-whistleblowers-letter/.

5 Kevin C. Limjoko, "The Philippines' Lost Opportunity," *Bugatti Review* 7, no. 4 (undated, c. 2007).

6 James Ritchie, "A wedding to remember," *New Straits Times*, August 4, 1991.

7 Sarawak CM's son may be charged for assault, *Malaysiakini*, April 30, 2003.

8 City-Data, "Property valuation of California Street, San Francisco," http://www. city-data. com /san-francisco/C/ California-Street-2. html.

9 Ross J Boyert vs. Sakti International Corporation Inc., Case no. CGC-07-460255, San Francisco Superior Courts, February 6, 2007.

10 King County, Washington, Recorders Office, Instruments no. OPR1224022 and OPR199112301455, December 26/27, 1991.

11 Webpage of the FBI Seattle, http://www. fbi.gov/seattle/about-us/what-we-investigate/priorities.

12 Ross Boyert, Diary, December 8, 1994, Boyert documents, Bruno Manser Fund archive, Basel.

13 Murray and Murray Associates Inc., MG 28, III 117, Finding Aid, Library and Archives Canada.

14 The Ashburian, *Yearbook of Ashbury College* (1982), 28, http://archive.org/details/ashburian-198200ashb.

15 Ibid. 41 and 174.

16 Elmwood School, *Report to the Community* (2010–2011), 24.

17 The Ashburian, *Yearbook of Ashbury College 1982*, http://

archive.org/details/
ashburian198200ashb.

18 The Sakto Property
Management
Corporation, which
was founded in 1987,
was renamed Orchid
Corporation in 1992
and City Gate Inter-
national Corporation in
1994. Source: Industry
Canada; Ontario
Ministry of Consumer
and Commercial
Relations.

19 In addition to the Sakto
Development
Corporation, the Sakto
group also owns the
following: Sakto
Corporation (founded
in 1997), Tower One
Holding Corporation,
Tower Two Holding
Corporation, Adelaide
Ottawa Corporation,
Sakto Management
Services Corporation,
Preston Building
Holding Corporation,
and 1041229 Ontario
Inc. All of these
companies are run by
Jamilah Taib and Sean
Murray. Source:
Industry Canada;
Ontario Ministry of
Consumer and
Commercial Relations.

20 In 2002, the Sakto
website stated for its
first year of operations,
"Incorporated in
August 1983, Sakto
acquired over 400

residential units."
http://web.archive.org/
web/20020208104337/;
http://sakto.com/
company.html. See also
Clare Rewcastle,
"Exclusive—Taib's
Foreign Property
Portfolio," *Sarawak
Report*, June 17, 2010.

21 The building at 333
Preston Street became
known as the Xerox
Building, http://www.
emporis.com/building/
xeroxbuilding-ottawa-
canada.

22 Sakto Development
Corporation, Balance
Sheet as at August 31,
1990. Microfiche
collection, Western
Libraries, Ontario.

23 Sakto Development
Corporation, Balance
Sheet as at August 31,
1992. Microfiche
collection, Western
Libraries, Ontario.

24 In 1996, for instance,
Sakto obtained a loan
worth 20 million
Canadian dollars from
the Taib family and its
companies, Richfold
Investment Ltd. in
Hong Kong and Sogo
Holdings Ltd. in
Jersey. Charge/
Mortgage of Land,
LT994558 and
LT994559, Ottawa-
Carleton Land Registry
Office, August 19,
1996.

25 Stephen Sigurdson,
Executive Vice
President and General
Counsel, Manulife
Financial, to the Bruno
Manser Fund, May 2,
2014; Instruments no.
OC903223,
OC903269,
OC903294,
OC248221, OC318707,
Ottawa-Carleton Land
Registry Office.

26 Kathrin May, "Pacific
Rim investment in
Canada," *Ottawa
Citizen*, January 17,
1989.

27 See also Clare
Rewcastle, "Taibs'
Lucrative Links with
Ontario Government,"
Sarawak Report, June
18, 2010.

28 In 2003, Murray &
Murray was acquired
by the IBI Group, an
architectural group
with worldwide
activities, employing
2,900 people in 79
countries.

29 Sarah Murray, Sean's
sister, is married to the
architect Nicholas
Caragianis. At the time
of writing, both work
together in the
architectural practice
of Nicholas Caragianis
Inc. in Ottawa. Its
references include
building the ornate
villa occupied by Sean
Murray and Jamilah

Taib in Rockcliffe Park, http://ncarchitect.ca.

30 Some details of the history of Laila Taib's family are to be found in chapter 4 of this book. Abu-Bekir Chalecki married for a second time in Ottawa and died there in 2004 at the age of 90. See also "Haji Chalecki: In memoriam," *Ottawa Citizen*, March 11, 2010.

31 In 2002, he assumed the presidency of CMS. Cahya Mata Sarawak, *Annual Report 2002*, 27.

32 The exchange rates used in this book are those of January 2014.

33 Utama Banking Group, *Annual Report 2005*, Profile of Directors, 3; Stephanie Phang, "RHB Chairman Fails to win Central Bank Approval for 2nd Term," *Bloomberg*, August 3, 2005. A record of the acquisition of the Utama Banking Group by the Taibs between 1991 and 1993 is to be found in Andrew Aeria, *Politics, Business, the State and Development in Sarawak 1970–2000* (Ph.D. thesis, University of London, 2002), 154 ff.

34 Action by Unanimous Written Consent of the Holders of all Outstanding Shares of Sakti International Holdings Inc., October 27, 2006, Boyert documents, Bruno Manser Fund archive.

35 Ross J Boyert vs. Sakti International Corporation Inc., Case Number CGC-07-460255, San Francisco Superior Courts, February 6, 2007.

36 "Grief, tears and death" was the title of an interview that Piero Grasso, Italy's chief mafia hunter, gave to Walter De Gregorio, in *Das Magazin*, no. 30&31 (2010): 20.

02 PARADISE LOST

1 The Penan headman Along Sega was interviewed by the author on July 8, 2005. The author would like to thank Ian Mackenzie for his comments on this interview and for the information on the Penan included in this chapter.

2 *Eugeissona utilis*, a member of the *Arecaceae* (palm) family. In addition to *uvut*, the Penan use six other sago palms containing starch, including *jakah* (*Arenga undulatifolia*).

3 *Antiaris toxicaria*, a member of the *Moraceae* (mulberry) family.

4 *Goniothalamus tapis*, a member of the *Annonaceae* (custard apple) family.

5 *Getimang* is mentioned as an antidote by Bruno Manser in his *Diaries from the Rainforest*. Ian Mackenzie has also encountered various other plants that are said to have the same or a similar effect. Mackenzie questions whether this is really true beyond a placebo effect. The author is not aware of any scientific research on this matter.

6 See also Ian Mackenzie, "The Eagle Augurs War," in *My Life as a Nomad, the Memoirs of Galang Ayu* (unpublished).

7 Alfred Russel Wallace, *The Malay Archipelago*, revised on the basis of the German translation of 1869 (Frankfurt, 1983), 36.

8 Ibid. 70.

9 Ibid. 55.

10 Ibid. 64.

11 *Rhacophorus nigropalmatus*

12 Alfred Russel Wallace, "On the law which has

regulated the introduction of new species," in *Annals and Magazine of Natural History* (September 1855). Cited from the website of the London Natural History Museum, http://www.nhm.ac.uk/nature-online/collections-at-the-museum/wallace-collection/closeup.jsp?itemID=138&theme=Evolution.

13 Alfred Russel Wallace, *Tropical Nature and other Essays* (London, 1878), 20 ff.

14 Charles Hose, *A Descriptive Account of the Mammals of Borneo* (London, 1893).

15 See also Charles Hose, *The Pagan Tribes of Borneo,* 2 vols (London, 1966).

16 Charles Hose, *Natural Man.* With an introduction by Brian Durrans. (London, 1987), vii.

17 Ibid. 39.

18 Charles Hose, "The Natives of Borneo," in *The Journal of the Anthropological Institute of Great Britain and Ireland* 23 (1894): 156–172, here 158.

19 Hose, *Pagan Tribes,* vol. 2, 180.

20 Hose, *Natives of Borneo,* 157 ff.

21 Ter Ellingson, *The Myth of the Noble Savage* (Berkeley, 2001).

There is a verse in Dryden's play *The Conquest of Granada* (1672) which runs: "I am as free as nature first made man, Ere the base laws of servitude began, When wild in woods the noble savage ran." As quoted in Ellingson, *Myth,* 36.

22 See also Jean-Jacques Rousseau, *Schriften zur Kulturkritik.* Introduced, translated and published by Kurt Weigand (Hamburg, 1983), 71, 79, and 89.

23 Hose, Natural Man, 1987, vii.

24 Tom Harrisson et al., eds., *Borneo Jungle: An account of the Oxford University Expedition of 1932* (Singapore, 1988).

25 The Canadian linguist and Penan researcher Ian Mackenzie points out that Needham's dissertation was actually a unique piece of evidence that the Western Penan and the Eastern Penan are two different peoples. Needham, however, failed to draw the logical conclusion from his observations.

26 This term was coined by the American anthropologist Peter Brosius.

27 Rodney Needham, *The social organisation of*

the Penan, a southeast Asian people (Ph.D., Oxford, 1953).

28 Needham writing to the author on September 7, 2006.

29 The following passages and the quotations from Ian Mackenzie refer to Andrew Gregg, "The Last Nomads" (ARTE/CBC, 2008), Documentary film, 53 min; and an interview with Ian Mackenzie in Basel.

03 THE WHITE RAJAHS

1 Steven Runciman, *The White Rajahs: A History of Sarawak from 1841 to 1946* (Cambridge, 2009) [Original 1960], 35 ff.

2 R.H.W. Reece, *The Name of Brooke: The End of White Rajah Rule in Sarawak* (Kuala Lumpur, 1982), 3.

3 Cf. Transfer by Pangeran Muda Hassim of the Government of Sarawak, 1841, and Appointment by Sultan of Brunei of James Brooke to Govern as His Representative, 1842, in Reece, *Name of Brooke,* 284 ff.

4 Faisal S. Hazis, *Domination and Contestation: Muslim Bumipu-*

*tera Politics in Sar-
awak* (Singapore,
2012), 28.

5   Runciman, *White
Rajahs*, 156.

6   He expressed explicit
criticism of imperial-
ism in his pamphlet
*Queries, Past, Present
and Future*, which ap-
peared in 1907.

7   Anthony Brooke, *The
Facts about Sarawak*
(Bombay, 1946), 32.
Cited in Reece, *Name
of Brooke*, 7.

8   The Australian histori-
an Bob Reece judged
this as follows: "The
justifying myth of
Brooke rule, as it had
evolved by the end of
the nineteenth century,
was the idea of trustee-
ship—that Sarawak
belonged to the people
and that the Rajah exer-
cised authority on their
behalf and in their
interests." Reece,
Name of Brooke, 11.

9   Reece, *Name of Brooke*,
98 ff.

10 Reece, *Name of Brooke*,
194.

11 Ibid. 200. To be fair to
Datu Patinggi, who was
the person concerned,
it ought to be added
that he was the only one
of the bribed Datus
who later paid the bribe
of 12,000 dollars back
to the Sarawak state.
Up to his death in

November 1946, he
became the symbolic
figure of the resistance
to the cession of Sara-
wak to the British
Crown.

12 Ibid. 236.

13 Ibid. 221.

14 "We found everywhere
affectionate loyalty for
the Brooke rule and
Rajah, whose personal
and intimate form of
government they
understand and appreci-
ate." Gammans, *Par-
liamentary Mission to
Sarawak* (June 1946),
18.

15 Reece, *Name of Brooke*,
267.

16 Ibid. 270.

17 Ibid. 276 ff.

18 An obituary appeared
in the New Zealand
Herald on March 3,
2011, and in the Tele-
graph on March 9.
Under the presidency
of Anthony's son,
James, the Brooke
Trust is today dedi-
cated to maintaining
the memory of the
Brooke dynasty, www.
brooketrust.org. Read-
ers are also referred to
the website set up by
Gita and Anthony
Brooke, http://www.
angelfire.com/journal/
brooke2000.

19 James Ritchie, *Temenggong Oyong Lawai Jau:
A Paramount chief in*

Borneo (Kuching,
2006), 43.

20 Speech of Temonggong
[sic] Oyong Lawai Jau,
MBE, QMC, January
1962, 24. Bodleian Li-
brary of Common-
wealth and African
Studies at Rhodes
House, Oxford, Mss,
109.

21 Ibid. 25

22 Ibid.

23 Ibid. 26.

24 Statement by Penghulu
Jok Ngau, Note on
[Cobbold] Commission
Hearing, Long Akah,
March 13, 1962. The
National Archives,
London.

25 Report of the Commis-
sion of Enquiry, *North
Borneo and Sarawak*
(London, 1962), 2.

26 Harun bin Abdul Ma-
jid, *The Brunei Rebel-
lion: December 1962;
The Popular Uprising*,
www.bruneiresources.
com/pdf/nd06_harun.
pdf. See also Alastair
Morrison, *Fair Land
Sarawak: Some Recol-
lections of an Expatri-
ate Official* (Ithaca,
1993), 141 ff.

27 In a confidential note to
Cabinet Secretary Nor-
man Brook in June
1962, Premier Macmil-
lan emphasised the
military weakness of
the British in Southeast
Asia and attached the

highest urgency in particular to transferring the security problems in Singapore to a new Malaysian state. Prime Minister's personal minute, no. 161/62, to Sir Norman Brook, June 21, 1962, The National Archives, London.

28 Ah Chon Ho, *Datuk Stephen Kalong Ningkan: First Chief Minister of Sarawak* (Kuching, 1992), 1.

04 SARAWAK'S MACHIAVELLI

1 Aeria, *Politics,* 164. See also Siva Kumar G, "The family's genealogy," in *Taib, a vision for Sarawak* (1991), xiv.

2 Kumar, *Taib,* 12.

3 Kylar Loussikian, "Student protest over Taib Mahmud Plaza in Adelaide," *Australian,* September 13, 2013. "YAB Datuk Patinggi Tan Sri (Dr) Haji Abdul Taib bin Mahmud," in *The Colombo Plan for cooperative economic development in South and South East Asia 1951–2001: The Malaysian-Australian Perspective.* A commemorative Volume to celebrate the 50th Anniversary of the Colombo Plan (Adelaide, 2001), 29.

4 "New court honours Chief Minister," *Adelaidean,* December 2008, http://www.adelaide.edu.au/adelaidean/issues/30821/news30825.html.

5 Australian Securities and Investments Commission (ASIC), Documents on Sitehost Pty Ltd, Australian Company Number 062312743, accessed March 25, 2010.

6 Laila (sometimes spelt Lejla) Chalecki (1941–2009) and her parents, Abu-Bekir Chalecki (1914–2004) and Dzemila Chalecki (Chalecka) (1916–1952), arrived in the port of Adelaide on November 22/23, 1949, onboard the *SS Oxfordshire*, a refugee ship organised by the International Refugee Organisation (IRO). It had set sail from Naples on October 22, 1949, http://www.immigrantships.net/v6/1900v6/oxfordshire19491123_01.html, accessed August 10, 2012. Documentation on the Chalecki family is also to be found in the National Archives of Australia.

7 For more information on the Lipka Tatars, who have been living for more than 600 years in Poland and Lithuania, see Jurgita Šiaučiūnaitė-Verbickienė, "The Tatars," in Grigorijus Potašenko, ed., *The Peoples of the Grand Duchy of Lithuania* (Vilnius, 2002), 73 ff.

8 Adas Jakubauskas, "Abu Bekiras Chaleckis (1914–2004)," Obituary in *Lietuvos totorių bendruomenio sąjungos laikraðtis* 74, no. 3 (2004).

9 Laila's mother died on February 25, 1952, in the Royal Adelaide Hospital. Death announcement in *Advertiser (Adelaide),* February 26, 1952.

10 On Hijjas Kasturi as a Colombo Plan scholar, see National Archives of Australia, A 1501, A2839/1.

11 See also photograph no. A 2840/1, series A 1501, National Archives of Australia.

12 Michael D. Leigh, *The Rising Moon: Political Change in Sarawak* (Sydney, 1974), 30 ff.

13 It is reported that later on Tunku complained that Rahman, whom he had built up politically, was insufficiently loyal

to him. See also Kumar, *Taib*, 4.

14 Leigh, *Rising Moon*, 31.

15 Vernon Porritt, *British Colonial Rule in Sarawak, 1946–1963* (Kuala Lumpur, 1997), 45 ff.

16 See also Alastair Morrison, *Fair Land Sarawak: Some recollections of an Expatriate Official* (Ithaca, 1993).

17 James Ritchie, *Abdul Taib Mahmud: 41 Years in the News* (Kuching 2005), 15.

18 Interview with a contemporary, Kuching, September 2012.

19 Mohammad Tufail Bin Mahmud, Cik Hanifah Taib, and Ritchie, James, eds., "Responsible Brother," in *Happy Wedding Anniversary Abang Taib and Kak Laila, 13 January 1999* (Kuching, 1999).

20 See also Morrison, *Fair Land Sarawak*, 147 ff.

21 Ho Ah Chon, *Datuk Stephen Kalong Ningkan, First Chief Minister of Sarawak* (Kuching, 1992). See also "Remembering Dad: Tan Sri Stephen Kalong Ningkan," *Borneo Post*, April 3, 2010, http://www.theborneopost.com/2010/04/03/remembering-dad-tan-sri-stephen-kalong-ningkan.

22 Kumar, *Taib*, 15.

23 The posts of state secretary, financial secretary and attorney-general were occupied by British expatriates. See also Leigh, *Rising Moon*, 83.

24 Ibid. 88–94.

25 Ho, *Stephen Kalong Ningkan*, 73.

26 Ritchie, *Abdul Taib Mahmud*, 34.

27 *Sarawak Tribune*, July 4, 1966, as quoted in Leigh, *Rising Moon*, 105.

28 Leigh, *Rising Moon*, 111.

29 What Taib said of Sli in 1991: "[He was] not tough enough to deal with a number of undercurrents he faced in Sarawak at the time." See also Kumar, *Taib*, 16.

30 R.S. Milne and K.J. Ratnam, *Malaysia— New States in a New Nation: Political Development of Sarawak and Sabah in Malaysia* (London: Frank Cass, 1974), 345.

31 Ibid.

32 *Sabah Times*, June 15, 1967, as quoted in Milne and Ratnam, *Malaysia*, 317.

33 David Walter Brown, *Why Governments Fail to Capture Economic Rent: The Unofficial Appropriation of Rain Forest Rent by Rulers in Insular Southeast Asia Between 1970 and 1999* (Ph.D. thesis, University of Washington, 2001), 313.

34 Michael L. Ross, *Timber Booms and Institutional Breakdown in Southeast Asia* (Cambridge, 2001), 64 and 71 ff.

35 Milne and Ratnam, *Malaysia*, 318.

36 See also James Wong, *The Price of Loyalty* (Singapore, 1983), 5.

37 Milne and Ratnam: *Malaysia*, 318; see also Leigh, *Rising Moon*, 116.

38 Kumar, *Taib*, 17.

39 *Vanguard*, October 9, 1967, as quoted in Leigh, *Rising Moon*, 115.

40 Leigh, *Rising Moon*, 115.

41 Milne and Ratnam, *Malaysia*, 318 and 329. See also Leigh, *Rising Moon*.

42 Ritchie, *Abdul Taib Mahmud*, 55.

43 Kumar, *Taib*. Foreword by Tunku Abdul Rahman Putra Al-Haj Bapa Malaysia. vii.

44 See also *Sarawak Tribune*, November 16, 1969, as quoted in Leigh, *Rising Moon*, 132.

45 The Miri oilfield reached its production peak in 1929. By 1945,

it had produced nearly 90% of its all-time oil output (80 million barrels). In 1972, production was closed down. See also Mario Wannier et al., *Geological Excursions around Miri, Sarawak* (Miri, 2011), 18.

46 Leigh, *Rising Moon*, 133.

47 Faisal S. Hazis, *Domination and Contestation: Muslim Bumiputera Politics in Sarawak* (Singapore, 2012), 92.

48 Ibid. 93.

49 Wee Chong Hui, *Sabah and Sarawak in the Malaysian economy* (Kuala Lumpur, 1995), 43. As cited in Hazis, *Domination and Contestation*, 93. The exchange rate at the beginning of 1980 was US$ 46 = MYR 100.

50 The LNG (liquid natural gas) plant in Bintulu was a joint venture between Petronas, Shell, and Mitsubishi. Most of its output went to Japan. See also Peter Hills, and Paddy Bowie, *China and Malaysia: Social and economic effects of petroleum development* (Geneva: International Labour Office, 1987), 104.

05 BLOWPIPES AGAINST BULLDOZERS

1 Ontario Ministry of Consumer and Commercial Relations, Records on Sakto Development Corporation.

2 ICRIS CSC Companies Registry, The Government of the Hong Kong Special Administrative Region, Regent Star Company Ltd, *Certificate of Incorporation* (November 22, 1983).

3 Same source: Regent Star Company Ltd., *Annual Return 1984*, and Richfold Investment Ltd., *Annual Return 1984*.

4 Bruno Manser Fund, *The Taib Timber Mafia: Facts and Figures on Politically Exposed Persons (PEPs) from Sarawak, Malaysia* (Basel, 2012), 16 ff.

5 Mark Baker, "Tycoon dodges millions in land tax," *The Age*, April 28, 2013.

6 Sarawak Timber Industry Development Corporation (1988), cited from Bruno Manser, *Voices from the Rainforest: Testimonies of a Threatened People* (Berne, 1992), 280.

7 Ross, *Timber Booms*, 146. The FAO is still keeping this study under lock and key and has refused the Bruno Manser Fund's request to see it.

8 K.S. Jomo et al., *Deforesting Malaysia: The Political Economy and Social Ecology of Agricultural Expansion and Commercial Logging* (London, 2004), 172.

9 Mission established pursuant to resolution I (VI), *The Promotion of Sustainable Forest Management: A Case Study in Sarawak, Malaysia*, Report Submitted to the International Tropical Timber Council (Bali, 1990), 71.

10 Forest Department Sarawak, Log Production And Forest Revenue 2000–2012, http://www.forestry.sarawak.gov.my/page.php?id=1030&menu_id=0&sub_id=28.

11 *New Straits Times*, April 10 and 12, 1987, cited from Aeria, *Politics*, 165 and 272 ff.

12 Aeria, Politics, 165.

13 Daniel Faeh, *Development of Timber Tycoons in Sarawak: History and Company Profiles* (Basel, 2009).

14 James Wong Kim Min has a variegated career behind him. He was already a minister in the first Sarawak cabinet in 1963, later

became a member of the federal parliament and for a short period was the parliamentary leader of the opposition. In October 1974, he was arrested on grounds of alleged conspiracy. He was later rehabilitated. See also James Kim Min Wong, *The Price of Loyalty* (Singapore, 1983).

15 Interview with Harrison Ngau Laing, Miri, 4 September 2012.

16 This protest is also described in Evelyne Hong, *Natives of Sarawak: Survival in Borneo's Vanishing Forests* (Pulau Pinang, 1987).

17 Manser, *Voices*, 263.

18 On "Operation Lalang," see also Barry Wain, *Malaysian Maverick. Mahathir Mohamad in Turbulent Times* (London, 2009), 65 ff.

19 Cited from Ruedi Suter, *Bruno Manser: Die Stimme des Waldes* (Berne, 2005), 35 ff.

20 Bruno Manser, *Diaries from the Rainforest,* Diary 12 (Basel, 2004), 149.

21 Rolf Bökemeier, "Ihr habt die Welt—lasst uns den Wald!" *GEO*, no. 10 (October, 1986), 12 ff; Suter, *Manser,* 147 ff.

22 Interview with Roger Graf, joint founder of the Bruno Manser Fund, in *Tong Tana,* magazine of the Bruno Manser Fund, December 2011.

23 See also Suter, *Manser,* 133 ff and 141 ff.

24 Faisal S. Hazis, *Domination and Contestation: Muslim Bumiputera Politics in Sarawak* (Singapore, 2012), 124.

25 Ibid. 132. See also the presentation from a somewhat different angle in Michael Leigh, "Money Politics and Dayak Nationalism: The 1987 Sarawak State Election," in Muhammad Ikmal Said and Johan Saravanamuttu, eds., *Images of Malaysia* (Kuala Lumpur, 1991).

26 Leigh, *Money Politics,* 191.

27 Ibid. 192.

28 Ibid. 194.

29 James Ritchie, *Sarawak: A Gentleman's Victory for Taib Mahmud* (Petaling Jaya, 1987).

30 See also Salleh Jaffaruddin, *Pricking Conscience* (unpublished manuscript, Kuala Lumpur, 2011), in particular chapter 3 dealing with the "Ming Court Affair."

31 Interview with a friend of Rahman Ya'kub's, September 2012.

32 Azman Ujang, "Truly memorable 80th birthday," *Star (Malaysia)*, January 1, 2008.

33 On the issue of Taib's bomohs, see Clare Rewcastle, "Taib's Secret Bomoh," *Sarawak Report*, October 23, 2010; Raziah's Blond Bomoh Bewitched Taib, in *Sarawak Report*, March 22, 2011.

34 Interview with Jürgen Blaser, Zollikofen, September 21, 2012.

35 See also Jürgen Blaser, *Transboundary Biodiversity Conservation: The Pulong Tau National Park, Sarawak State, Malaysia*, ITTO Project Supervisory Mission, 1–9 March 2006—PD 224/03, http://www.tropical-forests.ch/PD_224_03.php.

36 Aeria, *Politics*, 169 ff.

37 Ibid. 167.

38 Ibid. 170.

39 Ibid. 169.

40 Ibid. 173.

41 Ibid.

42 Ibid. 172.

43 See also Neilson Ilan Mersat, *Politics and Business in Sarawak (1963–2004)* (Ph.D., Canberra: Australian National University, 2005).

44 Manser, *Voices*, 197.

45 Suter, *Manser*, 180.

46 Website of the Registry of Societies Malaysia, Introduction to the Department, http://www.ros.gov.my/index.php/en/maklumat-korporat/pengenalan-jabatan.

47 Lim Kit Siang, "The re-arrest of Anderson Mutang Urud under the Emergency Ordinance after he was released on a court bail is a gross abuse of powers by the government and a blot on the human rights record of Malaysia." Opinion of the parliamentary opposition leader, Lim Kit Siang, March 4, 1992, http://bibliotheca.limkitsiang.com/1992/03/04/the-re-arrest-of-anderson-mutang-urud-under-the-emer%C2%ACgency-ordinance-after-he-was-released-on-a-court-bail-is-a-gross-abuse-of-powers-by-the-government-and-a-blot-on-the-human-rights-record-of-ma.

48 Letter from Dr. Mahathir Mohamad, Malaysian Prime Minister, to Bruno Manser, March 3, 1992. Bruno Manser Fund archive, Basel.

49 Wade Davis et al., *Nomads of the Dawn: The Penan of the Borneo Rainforest* (San Francisco, 1995), 140.

50 See also IDEAL (Institute for Development and Alternative Lifestyle), *Not Development, but Theft: The testimony of Penan communities in Sarawak* (Sibu, 2000), 9.

51 Ibid. 46 ff.

52 Ibid. 43–44.

53 Ruedi Suter, "Bruno Manser will sich in Malaysia stellen," *Online Reports*, April 7, 1998, http://archiv.onlinereports.ch/1998/Manser-Malaysia.htm

54 Interview with Roger Graf, joint founder of the Bruno Manser Fund, in *Tong Tana*, December 2011.

55 Ruedi Suter, "Das unerklärliche Verschwinden von Bruno Manser," *Online Reports*, November 19, 2000, http://archiv.onlinereports.ch/2000/ManserVermisstHintergrund.htm.

56 Peter Knechtli, "Der Erfolg in Sarawak ist unter Null," *Online Reports*, May 10, 1999, http://archiv.onlinereports.ch/1999/Manser-Augenzeuge.htm.

06 BRUNO MANSER'S LEGACY

1 At the end of 1997, the Supreme Court of Canada ruled that the Delgamuukw in British Columbia had a titular right to the land and not just rights of usufruct for fishing, hunting, etc. This case is seen as a turning point in the question of indigenous land rights. See also BC Treaty Commission: *Delgamuukw. A Lay Person's Guide to Delgamuukw* (Vancouver, 1999), www.bctreaty.net/files_3/pdf_documents/delgamuukw.pdf.

2 High Court (Kuching), "Nor Anak Nyawai & Ors v Borneo Pulp Plantation Sdn Bhd & Ors., suit no. 22-28 OF 1999-I," *Malayan Law Journal* 6 (2001): 241 ff.

3 "Justice Ian Chin tells of threats and indoctrination attempt," *Star (Malaysia)*, June 11, 2008.

4 Two decades later, an international expert body found the dismissal of Salleh Abas to have been unconstitutional. See also Jacqueline Ann Surin, "Eminent panel finds sacking of Salleh Abas and

two others 'unjustified'
and 'unconstitutional',"
Nutgraph, August 29,
2008.

5 "Sarawak doesn't rec-
ognise community
mapping," *New Straits
Times*, November 4,
2001.

6 UNDRIP (the United
Nations Declarations
on the Rights of Indi-
genous Peoples) was
adopted by the UN
General Assembly in
New York on Septem-
ber 13, 2007, after
many years of drawn-
out negotiations, www.
un.org/esa/socdev/
unpfii/documents/
DRIPS_en.pdf.

7 Bruno Manser Fund,
"Report slams Sarawak
logging on native
lands," media release,
April 15, 2010; Carol
Yong, *Logging in Sara-
wak and the Rights of
Sarawak's Indigenous
Communities*, a report
produced for JOAN-
GOHUTAN by
IDEAL, April 2010.

8 "Ian Urquhart, 1919–
2012," Obituary, *Tele-
graph*, September 20,
2012.

9 Ian Alexander Norfolk
Urquhart, "Teknonyms
of the Baram River,"
*Sarawak Museum Jour-
nal* VIII (1958): 383–
393; same author,
"Baram Teknonyms—

II," *Sarawak Museum
Journal* VIII (1958):
735–740; same author,
"Some Sarawak Kin-
ship Terms," *Sarawak
Museum Journal IX*
(1959): 33–46.

10 At the end of 2005, Ian
Urquhart signed an
affidavit for the benefit
of the Penan in a land
rights suit. A copy is to
be found in the Bruno
Manser Fund archive,
Basel.

11 Samling Global Limit-
ed, "Samling Global
Sets Record Straight on
Long Benalih Issue,"
media release, Septem-
ber 13, 2007.

12 TK Bilong Oyoi and
TK Kelesau Naan to
Mr. Leongchin Cheun,
Manager Kelesa Camp,
July 22, 2007, Bruno
Manser Fund archive,
Basel.

13 *Discovery of the
remains of Kelesau
Na'an (deceased),*
Penan Community
Report, Bruno Manser
Fund archive, Basel.

14 Richard Lloyd Parry,
"Jungle Tribal Leader
Kelesau Naan Took on
The Loggers: It May
Have Cost Him His
Life," *Times/Times
Online*, January 4,
2008.

15 SUARAM, "Circum-
stances surrounding
late Long Kerong

Headman's death re-
quire transparent and
accountable Police
probe," media release,
March 18, 2008, Bruno
Manser Fund archive,
Basel.

16 Hilary Chiew, "Violated by loggers: Teenage
schoolgirls have be-
come the latest target of
unscrupulous timber
workers," *Star (Malay-
sia)*, October 6, 2008.

17 Tony Thien, "Taib:
Stop the 'lies' about
Sarawak," *Malaysiaki-
ni*, October 9, 2008.

18 Desmond Davidson,
"Sarawak Deputy Chief
Minister dismisses
claims by Bruno Man-
ser Fund," *New Straits
Times*, September 24,
2008; and Philip Kiew,
"Jabu blames Bruno for
Penans' backward-
ness," *Borneo Post*,
December 11, 2008.

19 Puvaneswary Devin-
dran, "Assemblyman
questions credibility of
Bruno Manser Founda-
tion," *Borneo Post*,
November 13, 2008.

20 Companies Commis-
sion of Malaysia,
Kristal Harta Sdn Bhd,
Company No. 279832-
U; same source: Hanib
Corporation Sdn Bhd.,
Company No. 482186-
V; same source: Tri-
bune Press Sdn. Bhd,
Company No. 719095-

V, accessed September 2011. See also Bruno Manser Fund, *The Taib Timber Mafia: Facts and Figures on Politically Exposed Persons (PEPs) from Sarawak, Malaysia* (Basel, 2012), 32.

21 Companies Commission of Malaysia, Borneo Post Sdn Bhd, Company No. 35650-V, accessed September 15, 2011. Another person with great influence over the media is timber baron Tiong Hiew King, the founder of Rimbunan Hijau (see chapter 8). His Media-Chinese-International group owns the Chinese-language Malaysian daily *Sin Chew Daily* as well as other media in Malaysia, China, Hong Kong, Indonesia, Canada, and the USA. See also the company's website, www.mediachinese-group.com.

22 In December 2008, the minister thanked the Bruno Manser Fund for a report on this matter. Other offices of the Malaysian government did not reply to the correspondence from the Bruno Manser Fund. See also Bruno Manser Fund, "Confidential Report for the Royal

Malaysian Police, Sexual Investigation Unit," November 28, 2008; Ministry of Women, Family and Community Development to the Bruno Manser Fund, December 18, 2008.

23 The government's inquiry report appeared solely in the Malay language with the title "Laporan jawatankuasa bertindak peringkat kebangsaan bagi menyiasat dakwaan penderaan seksual terhadap wanita kaum penan di Sarawak," September 2009.

24 See also Sean Yoong, "The Associate Press: Malaysia gov't report: Loggers raped Borneo girls," *Washington Post online*, September 9, 2009.

25 Penan Support Group, FORUM-ASIA and Asian Indigenous Women's Network (AIWN), "A Wider Context of Sexual Exploitation of Penan Women and Girls in Middle and Ulu Baram, Sarawak, Malaysia." An Independent Fact-Finding Mission Report by the Penan Support Group, FORUM-ASIA and Asian Indigenous Women's Network (AIWN) (Petaling Jaya, 2010).

26 Instruction from Chin That Thong, General Manager Syarikat Samling Timber Sdn Bhd, to all Samling employees dated July 9, 2010. Bruno Manser Fund archive, Basel.

07 OFFSHORE BUSINESS

1 Beat Schmid, "Ökologen kritisieren CS," *Sonntagszeitung*, March 11, 2007.

2 Hong Kong Stock Exchange, "Samling Global Limited," February 23, 2007.

3 Ibid. "Samling Global Limited: IPO Allotment Results," March 6, 2007; Börse Berlin, "Announcement concerning the delisting von Samling Global Ltd.," June 15, 2012.

4 Hong Kong Stock Exchange, "Samling Strategic Corporation and Samling Global Limited: Proposal to privatise Samling Global Limited," April 30, 2012.

5 Norwegian Ministry of Finance, "Three companies excluded from the Government Pension Fund Global," press release, August 23, 2010, http://www.

regjeringen.no/en/dep/
fin/press-center/
Press-releases/2010/
three-companies-ex-
cluded-from-the-gov-
ern.html?id=612790.

6 Global Witness, *In the
future, there will be no
forests left* (London,
November 2012), 3.

7 Clare Rewcastle, "Sam-
ling at 'Epicentre' of
Sub-prime Crash!,"
Sarawak Report, Sep-
tember 3, 2010. The
reader is also referred
to the webpages of Sun-
chase Holdings, http://
www.sunchasehold-
ings.com/pages/fea-
tured_investments/
national_land_fund.
htm; and Mountain-
house, http://www.
mountainhouse.net/
development_team/
trimark.php.

8 Clare Rewcastle, "$1
Dollar Mansion?," *Sar-
awak Report*, August
25, 2010.

9 Companies Commis-
sion Malaysia, Perdana
ParkCity Sdn Bhd, No-
vember 6, 2012; Same
source: Perkapalan Da-
mai Timur Sdn Bhd,
October 4, 2011.

10 This account is based
on witness statements
by a whistle-blower
("Ling") and UBS
bank documents in the
possession of the Bruno
Manser Fund.

11 Clare Rewcastle, "Ma-
laysian Foreign Minis-
ter Named in MACC
Investigation into
Sabah Timber Corrup-
tion," *Sarawak Report*,
April 5, 2012.

12 Cf. Criminal-law com-
plaint lodged by the
Bruno Manser Fund
with Public Prosecu-
tor's Office III of Can-
ton Zurich against UBS
AG and person or per-
sons unknown, dated
May 25, 2012, Bruno
Manser Fund archive,
Basel.

13 Letter from the Office
of the Swiss Attorney
General to Monika
Roth (lawyer), dated
August 29, 2012,
Bruno Manser Fund
archive, Basel.

14 Clare Rewcastle,
"Barging into Profit—
The Chia Family Sue
Yayasan Sabah For RM
84.4 million In Monop-
oly Row," *Sarawak Re-
port*, April 19, 2012.

15 Tony Chan, "In Malay-
sia, Sarawak has a Cash
Register on the Port,"
Malaysiakini, Novem-
ber 14, 2007.

16 Pereira Goncalves et
al., *Justice for Forests:
Improving Criminal
Justice Efforts to Com-
bat Illegal Logging*,
World Bank study
(Washington DC,
2012), 1.

17 Clare Rewcastle,
"'Hold on Trust For
Aman'—More Devas-
tating Evidence from
The ICAC Investiga-
tion," *Sarawak Report*,
April 15, 2012.

18 Written statement by
UBS to Swiss national
TV's news show
"10vor10", March 5,
2014.

19 Letter written by the
Deutsche Bank Privat-
und Geschäftskunden
AG, Duisburg, to the
Bruno Manser Fund,
dated October 8, 2004,
giving notice of closure
of account no. (134)
1678556 00.

20 Letter written by the
Deutsche Bank Privat-
und Geschäftskunden
AG, Quality Manage-
ment, to the Bruno
Manser Fund, dated
January 6, 2005.

21 UBS Investment Bank,
Sarawak Corporate
Sukuk Inc., US$
350,000,000 Trust
Certificates due 2009.
Issue prospectus, De-
cember 17, 2004, 67.

22 "LFX market cap
reaches USD 12 billion
with the listing of Sar-
awak International In-
corporated Notes,"
LFX News (Labuan),
August 4, 2005.

23 UBS Investment Bank,
"Sarawak Corporate
Sukuk Inc.: US$

350,000,000 Trust Certificates due 2009," *Issue prospectus*, December 17, 2004. The loan was floated jointly with the UBS subsidiary Noriba in Bahrain, which was fully integrated in the group in 2006. Cf. "UBS absorbs Noriba in the Group," media release, March 14, 2006.

24 In 2011 and 2012, Goldman Sachs placed loans amounting to US$ 800 million each on behalf of the companies Equisar International and SSG Resources, both of which are under the control of the Sarawak government. Cf. "Goldman flexes its muscle on rare Malaysian bond," *IFR Asia* 705, July 9, 2011; Jonathan Rogers, "Sarawak defies weak market," *IFR Asia* 766, September 29, 2012.

25 Matt Wirz and Alex Frangos, "Goldman Sees Payoff in Malaysia Bet—Firm has pocketed over $200 Million from Bond Deals, but Also Provided Fuel in a Political Fight," *The Wall Street Journal,* May 1, 2013; Matt Wirz, "Goldman and Malaysia: BFF From Way Back," *The Wall*

Street Journal, May 3, 2013.

26 Tony Thien, "Sarawak CM defends state investment in troubled 1st Silicon," *Malaysiakini,* December 16, 2004.

27 Joseph Tawie, "Bad investment leaves Sarawak RM2.5 billion poorer," *Free Malaysia Today*, October 20, 2010.

28 Ibid. "Where has RM 11bil gone, Taib?," *Free Malaysia Today*, January 2, 2013; "We could have had more roads, schools, hospitals," *Free Malaysia Today*, January 3, 2013.

29 Democratic Action Party, *Democratising Sarawak's Economy: Sarawak DAP's Alternative Budget 2010* (Kuching, 2009), 4.

30 Dev Kar and Sarah Freitas, *Illicit Financial Flows From Developing Countries: 2001–2010,* Global Financial Integrity (Washington DC, 2012), 16.

31 In 2012, these were Raymond Yeoh, Deutsche Bank's chief country officer in Malaysia, Nilesh Navlanka of Deutsche Bank Singapore and the Italo-Australian Luigi Fortunato Ghirardello, former head of global finance

Asia-Pacific of Deutsche Bank Singapore; Companies Commission Malaysia, Company file on K & N Kenanga Holdings Berhad, September 8, 2011; K & N Kenanga Holdings Berhad, *Annual Report 2011.*

32 Deutsche Bank has a 27% direct holding through its subsidiary, Deutsche Asia Pacific Holdings, in Kenanga Deutsche Futures Sdn Bhd. It holds a further 10.4 % through its investment holdings in K & N Kenanga Holdings. Companies Commission of Malaysia, Company Records on Kenanga Deutsche Futures Sdn Bhd, accessed August 9, 2011.

33 K & N Kenanga Holdings Berhad, *Annual Report 2011,* 13.

34 K & N Kenanga Holdings, *Annual Report 2011,* 13.

35 K & N Kenanga Holdings, *Annual Report 2011,* 65.

36 San Francisco Superior Courts, Ross J Boyert vs. Sakti International Corporation, Complaint by Ross J Boyert, February 6, 2007, 3.

37 Sogo Holdings concluded a financial agreement with Sakto, Taib's daughter, Jami-

lah, and other family members on December 31, 1995. Land Registry Office, Ottawa, Charge/Mortgage of Land, LT994559, August 19, 1996.

38 Jersey Companies Registry, Annual Return of Sogo Holdings Limited, Company No. 43148, made up to January 1, 2013.

39 Bruno Manser Fund, "Anti-Money Laundering Authority investigating Deutsche Bank," media release, September 12, 2011.

40 Letter from Hartmut Koschyk, Parliamentary State Secretary at the Federal Ministry of Finances, to Dr. Thomas Gambke (member of the Bundestag) dated March 8, 2012, Bruno Manser Fund archive, Basel.

41 Email from Deutsche Bank Privat- & Geschäftskunden AG, Vertriebs Region Südbaden, to the author, dated May 27, 2013, Bruno Manser Fund archive, Basel.

42 Letter from the Bruno Manser Fund to the Swiss Federal President, Micheline Calmy-Rey, dated March 17, 2011.

43 Letter from the Swiss Federal President, Mi-

cheline Calmy-Rey, to the Bruno Manser Fund, dated April 8, 2011.

44 Ang Ngan Toh, "Taib: I have no secret bank account," *Malaysiakini,* June 22, 2011.

45 Bernama, "Taib under MACC probe," June 9, 2011.

46 Hafiz Yatim, "Taib is richest person in Malaysia, says Shahnaz," *Malaysiakini*, October 2, 2012.

47 Shahnaz was a member of the CMS board from 1995 until January 2004 and deputy of Taib's brother Onn Mahmud. Cahya Mata Sarawak, *Annual reports* (1995 to 2004).

48 Ibid.

49 Media release by the Office of the Swiss Attorney General, September 24, 2013; see also the letter from the Office of the Swiss Attorney General to Carlo Sommaruga and the Bruno Manser Fund dated September 23, 2013, Bruno Manser Fund Archive, Basel.

08 TRAIL OF DESTRUCTION

1 Bruno Manser Fund and Society for Threat-

ened Peoples, *Credit Suisse asked to pay back profits of Samling listing*. Media kit compiled on the occasion of the media conference of May 3, 2007, http://www.bmf.ch/en/news/?show=51. See also Janette Bulkan and John Palmer, *Lazy days at international banks: How Credit Suisse and HSBC support illegal logging and unsustainable timber harvesting by Samling/Barama in Guyana, and possible reforms*. Report to Chatham House (Royal Institution for International Affairs, London, UK), FLEGT update meeting, July 10, 2007, http://www.illegal-logging.info/uploads/Samling_Barama.pdf.

2 Report by *Toshao* David Wilson at the meeting with Credit Suisse, Samling and an NGO delegation in Zurich on May 3, 2007; David Wilson, "Akawini Village calls for support to end 'bad faith' agreement with Barama," speech made to a media conference in Zurich, May 3, 2007.

3 See also Wikipedia overview of Guyana at http://en.wikipedia.org/wiki/Guyana.

4 The remaining Guy-
anese are declaring
themselves as Mixed
ethnicity. See Guyana
Bureau of Statistics,
*2002 Population and
Housing Census*, Guy-
ana National Report,
http://www.statistics-
guyana.gov.gy/census.
html.
5 Statement by Janette
Bulkan at the meeting
with Credit Suisse,
Samling, and an NGO
delegation in Zurich on
May 3, 2007.
6 Janette Bulkan and
John Palmer, *Illegal
logging by Asian-owned
enterprises in Guyana,
South America.* Brief-
ing paper for Forest
Trends' Second Poto-
mac Forum on Illegal
Logging & Associated
Trade (Washington
DC, February 14,
2008), 5. By 2013,
Asian control of me-
dium- and large-scale
logging concessions
had increased to 79 per
cent. Bulkan, 2014, in
press.
7 Janette Bulkan, "Fail-
ures by Credit Suisse to
implement its own
commitments," contri-
bution to the media
conference of the
Bruno Manser Fund
and the Society for
Threatened Peoples in
Zurich, May 3, 2007.

8 WWF, "Barama and
WWF to Influence
Global Markets
through Responsible
Forest Management in
South America," news
release, March 27,
2006.
9 "FSC audit of SGS
leads to suspension of
largest tropical forest
logging certificate,"
FSC Watch, January
18, 2007, http://www.
fsc-watch.org/arch-
ives/2007/01/18/FSC_
audit_of_SGS_leads_
to_suspension_of_
largest_tropical_log-
ging_certificate.
10 Tusika Martin,
"Akawini forces Bara-
ma to withdraw from
concession," *Kaieteur
News,* May 30, 2007.
11 Bruno Manser Fund,
"Guyana's Head of
State condemns Sam-
ling," media release,
October 19, 2007,
http://www.bmf.ch/en/
news/?show=77.
12 Bulkan and Palmer,
Illegal logging, 7.
13 Ibid. 8.
14 "WWF has 'discon-
nected' from Barama,"
Stabroek News, January
11, 2009.
15 On Sarawak as a "glo-
bal hotspot of tropical
deforestation" and the
role of the timber cor-
porations from Sara-
wak, see Jane Bryan et

al., "Extreme Differ-
ences in Forest Degrad-
ation in Borneo: Com-
paring Practices in Sa-
rawak, Sabah, and
Brunei," *PLoS ONE* 8,
no. 7, (2013),
doi:10.1371/journal.
pone.0069679.
16 Global Witness, *In the
future, there will be no
forest left* (London,
2012), 7.
17 Cf. website of the Rimbunan Hijau group,
http://www.rhg.com.
my.
18 See also Alain
Karsenty, *Overview of
Industrial Forest Con-
cessions and Conces-
sion-based Industry in
Central and West Af-
rica* (Montpellier,
2007).
19 See also Forest Trends,
*Logging, Legality and
Livelihoods in Papua
New Guinea: Synthesis
of Official Assessment of
the Large-Scale Log-
ging Industry* (2006),
10 ff, http://www.for-
est-trends.org/docu-
ments/files/doc_105.
pdf.
20 "Malaysia's 40
Richest," *Forbes*, 2010,
http://www.forbes.
com/lists/2010/84/
malaysia-rich-10_Yaw-
Teck-Seng-Yaw-Chee-
Ming_GSVV.html;
"Singapore's 40
Richest," *Forbes,* 2010,

http://www.forbes.
com/lists/2010/79/
singapore-10_Yaw-
Chee-Siew_WX0Y.
html); Ibid. http://
www.forbes.com/
lists/2012/84/
malaysia-billion-
aires-12_Abdul-
Hamed-Sepawi_N73D.
html.

21 Bruno Manser Fund,
*The Taib Timber
Mafia: Facts and Fig-
ures on Politically Ex-
posed Persons (PEPs)
from Sarawak, Malay-
sia* (Basel, 2012)

22 Steven Runciman, *The
White Rajahs*, (Cam-
bridge, 2009) [Ori-
ginal 1960], 208 ff.

23 Ibid.

24 "The great Foochow
factor," *New Straits
Times*, March 21, 2011.
The companies headed
by Fuzhou Chinese are
KTS, WTK, Rimbu-
nan Hijau, Shin Yang,
and Ta Ann. By con-
trast, the owners of
Samling, the Yaw fam-
ily, hailed originally
from Canton, China.

25 Rimbunan Hijau's web-
site, http://www.rhg.
com.my/about/early-
years.html.

26 After his release, James
Wong, under his full
name James Wong Kim
Min, wrote a book in
which he reported on
his experience. James

Wong Kim Min, *The
Price of Loyalty* (Singa-
pore, 1983).

27 This was written by
journalist SK Lau from
Sibu in his manuscript
"Immortal-Tiger-Dog"
published in the mid-
1990s. See also David
Walter Brown, *Why
Governments Fail to
Capture Economic
Rent: The Unofficial
Appropriation of Rain
Forest Rent by Rulers in
Insular Southeast Asia
Between 1970 and 1999*
(PhD, University of
Washington, 2001),
160.

28 Ibid. 161 ff.

29 Assuming a harvest of,
for instance, 80 cubic
metres of timber per
hectare and net pro-
ceeds of US$ 85 per
cubic metre (for costs
of roughly US$ 45 per
cubic metre), it is pos-
sible to make a profit of
US$ 10.2 billion from
exploiting 1.5 million
hectares of virgin forest
in Sarawak.

30 Joe Studwell, *Asian
Godfathers. Money and
Power in Hong Kong &
South-East Asia* (Lon-
don, 2008), 50 ff.

31 Wong Meng Chuo to
the author, January 23,
2013, Bruno Manser
Fund archive, Basel.

32 Alfred Russel Wallace,
*Der Malayische Archi-

pel* (Frankfurt, 1983)
[revised on the basis of
the German translation
of 1869], 402.

33 WWF, *Final Frontier:
Newly discovered species
of New Guinea (1998–
2008)* (WWF Western
Melanesia Programme
Office, 2011).

34 Thomas E. Barnett,
*The Barnett Report: a
summary of the report
of the Commission of
Inquiry into aspects of
the timber industry in
Papua New Guinea*
(Asia-Pacific Action
Group, 1990). On the
situation at the time of
writing, cf. Transpar-
ency International,
*Forest Governance In-
tegrity Baseline Report
Papua New Guinea*
(2011).

35 James Chin, "Contem-
porary Chinese Com-
munity in Papua–New
Guinea: Old Money
versus New Migrants,"
*Chinese Southern Dias-
pora Studies* 2 (2008):
120 ff.

36 Greenpeace, *Rimbu-
nan Hijau Group:
Thirty Years of Forest
Plunder* (Amsterdam,
2006), 3.

37 Greenpeace, *Partners
in Crime. Malaysian
loggers, timber markets
and the politics of
self-interest in Papua*

New Guinea (Amsterdam, 2002).

38 Radio Australia, *Big win against illegal logging in PNG*, June 27, 2011, http://www.radioaustralia.net.au/international/radio/onairhighlights/big-win-against-illegal-logging-in-png.

39 Greenpeace, *Rimbunan Hijau*, 4 ff.

40 World Bank, "Weak Forest Governance Costs us over US$ 15 Billion A Year," news release, no. 2007/86/SDN, September 16, 2006.

41 Greenpeace, *Rimbunan Hijau*, 4 ff.

42 John Vidal, "Forest campaigners deplore knighthood for Asian logging magnate," *Guardian*, July 1, 2009.

43 Website of Rimbunan Hijau: Milestones—Overview, http://www.rhg.com.my/about/mstone.html.

44 See also Alain Karsenty, *Overview*, 45; Bénédicte Chatel, "Bois tropicaux et conflit sur les terres: la Malaisie un exemple pour l'Afrique?" *Les Afriques*, no. 185, (January 12, 2012): 12.

45 International Union for Conservation of Nature (IUCN) and China

Wood International Inc., *Scoping study of the China-Africa timber trading chain* (Beijing, 2009).

46 See also Peter Maass, "Who's Africa's Worst Dictator?," *Slate online magazine* (www.slate.com), June 24, 2008; United States State Department, *2008 Human Rights Report: Equatorial Guinea,* http://www.state.gov/j/drl/rls/hrrpt/2008/af/118999.htm.

47 "Teodorin Obiang, le fils gâté qui siphonne son pays," *Courrier International,* April 12, 2012. In a spectacular action in 2011, the French police confiscated various luxury limousines belonging to Teodorin in Paris, after French NGOs had filed a complaint against Obiang. Xavier Harel and Thomas Hofnung, "Le Scandale des biens mal acquis: Enquête sur les milliards volés de la Françafrique" (Paris, 2011), 7 ff.

48 In 2012, the US Justice Department filed a case against "Teodorin" Obiang and had his assets in the USA confiscated. Cf. United States of America vs. One White Crys-

tal-Covered 'Bad Tour' Glove and others, Second Amended Verified complaint for Forfeiture in rem, US District Court for the Central District of California, June 11, 2012.

49 Karsenty, *Overview,* 18.

50 Emeric Billard, "Nouveaux acteurs, vieilles habitudes: L'implantation des opérateurs forestiers asiatiques au Gabon à l'heure de la transition vers la gestion durable." Thèse pour obtenir le grade de docteur du Muséum national d'Histoire naturelle (Ph.D. thesis, Paris, 2012), 43; Karsenty, *Overview,* 14 ff.

51 However there seems to have been a major policy change in Gabon after a new president took office.

52 Kerstin Canby et al., *Forest Products Trade between China & Africa: An Analysis of Imports and Exports,* published by Forest Trends and Global Timber (London, 2008), 21 ff.

53 Samling Global Limited, *Global Offering, Global Coordinator: Credit Suisse (Hong Kong) Limited,* (Hong Kong, February 23, 2007), 115 ff.

54 Brown, *Governments.*

55 On Samling in Cambodia see Global Witness, *The Untouchables: Forest crimes and the concessionaires – can Cambodia afford to keep them?,* briefing document, (December 1999), 10 ff.; same source: *Just Deserts for Cambodia,* 1997.

56 Global Witness, "Logging anarchy continues despite border closure," news release, March 3, 1997.

57 See also Global Witness, *Untouchables,* 10 ff.

58 Global Witness "SL International guilty of illegal forest exploitation—official," media release, May 23, 1997.

59 Asian Development Bank Sustainable Forest Management Project, *Cambodian Forest Concession Review Report* (2000).

60 Samling Global statement on Cambodia involvement, Bruno Manser Fund archive, Basel.

61 Trixie Carter, "Logging company told to pack up and leave," *Solomon Star,* December 29, 2009.

62 www.observertree.org

63 www.taann.com.my/corporate-profile

64 At the end of 2011, Sepawi held 9.37% of the shares in the parent company, Ta Ann Holdings. He held a further 26.1% of Ta Ann shares through indirect investments. Ta Ann Holdings Berhad, *Annual Report 2011,* 213 ff.

65 There is evidence showing that Taib used a construction like this for his US real estate holding, Sakti International (see chapter 1). The same presumption applies to Wahab Dolah, a member of Taib's political party and another major Ta Ann shareholder.

66 Bruno Manser Fund, *Taib Timber Mafia,* 24 ff.

67 Global Witness, *Pandering to the loggers: Why WWF's Global Forest and Trade Network isn't working* (July 2011), 8–11.

68 http://www.taann.com.my/bs-timber.html, and http://www.taann.com.my/bs-reforestation.html

69 Huon Valley Environment Centre, *Behind the Veneer: Forest Destruction and Ta Ann Tasmania's lies,"* (September 2011), 10 and 25. Over the first five years, Ta Ann Tasmania received direct subsidies out of tax money amounting to 10.3 million Australian dollars (US$ 9.4 million). To this must be added indirect subsidies totalling 23 million Australian dollars (US$ 21.7 million) paid to a state-owned company operating the Ta Ann sawmills.

70 Nick Clark, "Ta Ann in $10m loss," *Mercury,* July 20, 2013; Notorious Sarawak Timber Company receives Australian taxpayer handout, news release of the Bob Brown Foundation, May 9, 2014.

71 "Hydro Tasmania to Withdraw from Sarawak Dam-Building Program," *Environment News Service,* December 5, 2012.

72 Tony Burke, Simon Crean, and Joe Ludwig, "Supporting the Tasmanian Forestry Agreement," joint news release by the Hon Tony Burke MP, the Hon Simon Crean MP, and Senator the Hon Joe Ludwig, January 31, 2013.

73 http://observertree.org/2013/03/07/bushfire-forces-exit-from-observertree-mirandas-epic-tree-sit; Hannah Martin, "Mi-

randa's epic tree-sit ends," *Mercury (online edition)*, March 7, 2013.

74 http://observertree. org/2013/06/24/ media-release-the- world-celebrates-the- success-of-community- action-to-protect- forests; cf. http://whc. unesco.org/en/list/181

75 Andrew Darby, "Tony Abbott's bid to delist Tasmania's world heri- tage forests tipped to fail," *Sydney Morning Herald*, February 4, 2014; "Logging on: Tony Abbott reignites an environmental bat- tle," *Economist*, March 22, 2014. See also: ABC News: "UNESCO Rejects Coalition's Bid to Delist Tasmanian World Heritage Forest", June 24, 2014.

76 Greenpeace, *Logging the Planet: Asian Com- panies marching across our last forest frontiers* (May 1997), Green- peace submission to: External commission about foreign logging companies in the Ama- zon. An overview of Asian companies, in particular Malaysian companies.

77 Ibid. 4 ff.

78 Forests Monitor and World Rainforest Movement, *High Stakes: The Need to Control Transnational Logging Companies; a Malaysian Case Study* (August 1998).

79 Figueiredo Tautz and Carlos Sergio, "The Asian Invasion. Asian Multinationals Come to the Amazon," *Multi- national Monitor* 18 (1997). See also Green- peace, *WTK and the Deni: A Malaysian log- ging giant and indi- genous people in the Amazon* (2003).

09 GREEN WASTELAND

1 See also Stephen Then, "Plantation workers nabbed for staging hi- jacking, robbery," *Star (Malaysia)*, April 19, 2009.

2 Emily B. Fitzherbert et al., "How will oil palm expansion affect biodi- versity?," *Trends in Ecology and Evolution* 23 (2008): 529 ff.

3 Ibid.

4 The rule applicable in Switzerland, for in- stance, is that since March 2008 edible vegetable oil must not contain more than 2% of trans-fatty acids. In- formation provided by the Swiss Federal Of- fice of Public Health (FOPH) on May 28, 2013. See also Migros- Genossenschafts-Bund, "Migros: Gebäck mit weniger als 2 Prozent Transfettsäure," news release, September 6, 2007; Andreas Grämiger, "Ungesund und sehr gut versteckt," *saldo*, no. 9, May 12, 2004.

5 United States Depart- ment of Agriculture, Economic Research Service, World vege- table oils supply and distribution.

6 The Rainforest Foun- dation, *Seeds of De- struction: Expansion of industrial oil palm in the Congo Basin; Poten- tial impacts on forests and people* (London, 2013).

7 Fitzherbert et al., *Oil palm.*

8 Jan Willem van Gelder, *Greasy palms—Euro- pean buyers of Indones- ian palm oil*, (Friends of the Earth, 2004).

9 Marcus Colchester, Thomas Jalong, and Wong Meng Chuo, "Sarawak: IOI Pelita and the community of Long Teran Kanan," pre-publication text for public release, Septem- ber 2012, 9 ff.

10 Statement by Baya Sigah of Long Teran Kanan, cited in Col-

chester, Jalong, and Wong, 2012, 9.

11 "Malaysia's 40 richest," *Forbes,* 2012, http://www.forbes.com/lists/2012/84/malaysia-billionaires-12_Lee-Shin-Cheng_HZCA.html.

12 Cf. RSPO website, http://www.rspo.org.

13 Angus Stickler, "Borneo tribes 'driven from land'," *BBC News,* http://news.bbc.co.uk/2/hi/asia-pacific/8424156.stm.

14 Milieudefensie, and Friends of the Earth International, *Too Green to be True: IOI Corporation in Ketapang District, West Kalimantan* (March 2010).

15 Roundtable on Sustainable Palm Oil (RSPO), "Announcement on IOI by RSPO Grievance Panel: Breach of RSPO Code of Conduct 2.3 & Certification Systems 4.2.4c," April 5, 2011, http://www.rspo.org/news_details.php?nid=34&lang=1.

16 "IOI in talks to buy Achi Jaya for RM 800mil to RM900mil," *Malaysiakini,* April 17, 2013; Intan Farhana Zainul, "IOI Corp brushes off acquisition rumours," *Star (Malaysia),* June 14, 2013.

17 Personal communication from Matthias Diemer, head of the international projects department at WWF Switzerland, March 13, 2013.

18 Lian Pin Koh and David S. Wilcove, "Is oil palm agriculture really destroying tropical biodiversity?," *Conservation Letter* 1 (2008): 60 ff.

19 Swiss-Impex, import statistics published by the Swiss Directorate General of Customs, customs tariff position 1511, 2004 to 2012.

20 Rhett A. Butler, "E.U. OKs biofuels produced from certified palm oil," *mongabay (online),* November 28, 2012.

21 Neste Oil Corporation, "Neste Oil commits to using solely certified palm oil by the end of 2015," news release, June 4, 2009.

22 Just the refinery that started to operate in Singapore in November 2010 costs US$ 750 million. Cf. Neste Oil Corporation, "Singapore renewable diesel refinery," http://www.nesteoil.com/default.asp?path=1,41,537,2397,14090; same source: "Rotterdam renewable diesel

refinery," http://www.nesteoil.com/default.asp?path=1,41,537,2397,14089.

23 Neste Oil Corporation, *Annual Report 2012,* 128.

24 Rainforest Foundation Norway, "World's largest sovereign wealth fund divests from palm oil companies," news release, March 15, 2013.

25 See also Clare Rewcastle, "We release the land grab data!," *Sarawak Report,* March 19, 2011.

26 Bruno Manser Fund, "Oil palm plantation land leased to Taib linked companies in Sarawak," January 19, 2012, http://www.stop-timber-corruption.org/resources/Mapping_Taib_s_Land_Grabs___NEW_Blatt1_1.pdf; same source: "Summaries of Companies linked to Taib which have been leased palm oil land in Sarawak," http://www.stop-timber-corruption.org/resources/Companies_linked_to_Taib_which_have_been_leased_palm_oil.pdf.

27 Clare Rewcastle, "Deepening Scandal— Taib's Land Grabs Exposed!," *Sarawak Re-*

port, December 3, 2010, http://www.sarawakreport.org/2010/12/deepening-scandal-taibs-land-grabs-exposed/.

28 Global Witness, "Inside Malaysia's Shadow State," film (16:25 minutes), 2012, http://www.malaysiashadowstate.org, and http://www.youtube.com/watch?v=_1RRNggn-M6A.

29 Ample Agro is a 100% subsidiary of Sateras Holdings, which belongs jointly to six Rahman daughters. Companies Commission of Malaysia, files on Ample Agro Sdn Bhd, Company Number 821926-T; same source: Sateras Holdings Sdn Bhd, Company Number 67886-D, accessed March 19, 2013.

30 http://www.malaysiashadowstate.org, film, minute 05:55.

31 Bruno Manser Fund, "Malaysian authorities urged to close Taib's land-grab firms," media release, March 25, 2013.

32 Bruno Manser Fund, *Sold Down the River. How Sarawak Dam Plans Compromise the Future of Malaysia's Indigenous Peoples*

(Basel, September 2012), 12.

33 Cf. the official website of the dam, www.bakundam.com.

34 Benjamin K. Sovacool, and L.C. Bulan, "Meeting Targets, Missing People: The Energy Security Implications of the Sarawak Corridor of Renewable Energy (SCORE)," *Contemporary Southeast Asia* 33 (2011): 56 ff.

35 Bruno Manser Fund, "Leaked document details of Sarawak's excessive hydropower plans," media release, June 2008, http://www.bmf.ch/en/news/?show=101; Sarawak Energy, "Hydropower Projects in Sarawak 2008–2020," China-ASEAN Power Corporation & Development Forum, October 28–29, 2007, Nanning, China.

36 Agence France Press (AFP), "Malaysian government approves 700 kilometre undersea cable," April 26, 2009, http://www.bmf.ch/en/news/?show=147.

37 Bruno Manser Fund, *Sold Down the River*, 17 and 28.

38 Bruno Manser Fund, *Complicit in Corruption: Taib Mahmud's*

Norwegian Power Man, (Basel, May 2013), 16.

39 Bruno Manser Fund, "Sarawak Dams to Flood 2,300 km² of Rainforests, Displace Tens of Thousands of Natives," media release, May 17, 2013, http://www.bmf.ch/en/news/?show=344.

40 Jack Wong, "SEB plans another five power plants," *Star (Malaysia)*, October 18, 2012.

41 Cf. Priority Sector "Oil-based Industries" on the official SCORE website, http://www.recoda.com.my/priority-sectors/oil-based-industries.

42 100% of the shares in Sarawak Energy are held by the Sarawak ministry of finances. Companies Commission Malaysia, documents pertaining to Sarawak Energy, September 8, 2011.

43 Bruno Manser Fund, *Sarawak Dams*.

44 Torstein Dale Sjøtveit, "My Hometown, Rjukan," April 16, 2013, http://sarawakenergy.wordpress.com/2013/04/16/my-hometown-rjukan.

45 Torstein Dale Sjøtveit, "Sarawak Energy Torstein Dale Sjøtveit Long Wat Murum 4," March 25, 2013, http://

sarawakenergy.word-press.com/2013/03/26/my-visit-to-long-wat/sarawak-energy-tor-stein-dale-sjotveit-long-wat-murum-4.

46 Email message from Torstein Dale Sjøtveit dated May 14, 2013, Bruno Manser Fund archive, Basel.

47 Sarawak Energy, "Work Resumes at Murum Hydroelectric Plant," news release, September 30, 2013, http://www.sarawaken-ergy.com/index.php/news-events-top/latest-news-events/latest-media-release/461-work-resumes-at-murum-hydroelectric-plant.

48 Ibid.

49 Bruno Manser Fund, *Complicit in Corruption.*

50 *Sun Daily*, October 30, 2013.

51 "Torstein receives honorary 'Datuk'," *New Sarawak Tribune*, September 15, 2013.

52 On the state visit by Albert II, Principauté de Monaco, "H.S.H. Prince Albert II's visit to Asia," *news release*, April 2008, http://www.presse.gouv.mc/304/wwwnew.nsf/1909$/40ba823445829043c-125742d002c5e2eg-b?OpenDocument

&2Gb; Clare Rew-castle, "Billions Abroad!—Questions about Taib's Contacts with Foreign Property Tycoons," *Sarawak Report*, January 27, 2011; Kallakis was en-nobled by Albert II in November 2008 and accepted into the "Or-dre de Grimaldi".

53 Simon Bowers, "Bogus Mayfair property ty-coon convicted of £750m fraud," *Guard-ian*, January 16, 2013; Serious Fraud Office, "Achilleas Kallakis and Alexander Williams jailed," news release, January 17, 2013.

54 The Gentas had, inter alia, been guests at the marriage of Taib's niece, Elia Geneid; "Elia's Wedding," *Star (Malaysia)*, November 11, 2007.

55 Islamic Fashion Festi-val 2010, Monaco, Au-gust 9, 2010, http://www.youtube.com/watch?v=3O4n-Juoh4_M; Evelyne Genta was honoured with the noble title of "Datuk" by the Malay-sian king in 2011.

56 Robyn Mills, "New court honours Chief Minister," *Adelaidean*, December 2008, http://www.adelaide.edu.au/adelaidean/

issues/30821/news30825.html.

57 Joseph Tawie, "Sara-wak defying court on NCR, says PKR," *Free Malaysia Today*, November 26, 2012.

58 Friends of the Earth, *Malaysian palm oil—green gold or green wash? A commentary on the sustainability claims of Malaysia's palm oil lobby, with a special focus on the state of Sarawak* (October 2008), 38 and 54 ff.

59 Bruno Manser Fund, "Sarawak's natives must stand up for their rights and fight for them," Interview with PKR president Baru Bian, January 2010, http://www.bmf.ch/en/news/?show=193.

60 SAVE Rivers, About Sarawak Dams, http://www.savesarawak-rivers.com/about-the-dams.

61 See also Interview with Peter Kallang in *Tong Tana*, magazine of the Bruno Manser Fund, March 2012.

62 Environment News Service, "Sarawak Na-tive Leader Barred from Hydropower World Congress," May 20, 2013, http://ens-newswire.com/2013/05/20/sarawak-native-lead-

er-barred-from-hydro-power-world-congress.

63 International Institute for Sustainable Development, "Summary of the Fourth International Hydropower Association World Congress on advancing sustainable hydropower, 21–24 May 2013," http://www.iisd.ca/hydro/iha2013/html/crsvol-139num10e.html.

64 Lian Cheng, "Group protests building of more mega dams," Borneo Post, May 23, 2013, http://www.theborneopost.com/2013/05/23/group-protests-building-of-more-mega-dams.

65 Cf. the conference centre's website under Board of Directors, http://www.bcck.com.my/about.

66 Agence France Presse, "Outrage grows over scandal-tainted Malaysian leader," South China Morning Post, May 23, 2013.

10 RAINFORESTS WITHOUT CORRUPTION

1 For more information see www.bmf.ch and the campaign websites of www.stop-timber-corruption.org and www.stop-corruption-dams.org.

2 The broadcast can be watched on YouTube at http://www.youtube.com/watch?v=JObNkSds3lA; See also: the Global Television website: http://globalnews.ca/news/185008/family-trees-2.

3 Bruno Manser Fund, The Taib Timber Mafia: Facts and Figures on Politically Exposed Persons from Sarawak, Malaysia, (Basel, 2012). Further information in (inter alia) "Groundbreaking study details Taib's US$ 21 bil empire," Malaysiakini, September 19, 2012; Pushparani Thilaganthan, "Taib is worth RM45b, believe it or not," Free Malaysia Today, September 19, 2012.

4 See also www.sarawakreport.org, and www.radiofreesarawak.org.

5 International Press Institute, "IPI hands 2013 awards to two women journalists killed in Syria and independent Malaysian radio station," news release, May 2, 2013.

6 Malaysia's position in the international Press Freedom Index is a lowly 145th. Reporters without borders, Press Freedom Index 2013, http://en.rsf.org/spip.php?page=classement&id_rubrique=1054.

7 "Taib Mahmud being investigated, says MACC," Star (Malaysia), June 9, 2011.

8 Gerry Mullany, "Malaysia Denies Entry to Journalist," New York Times, July 4, 2013.

9 See also "Masing's wife received millions in contracts," Free Malaysia Today, March 13, 2012.

10 Statement by Ketua Menteri Sarawak (Y.A.B. Pehin Sri Haji Abdul Taib bin Mahmud), Hansard of the Sarawak State Assembly, May 29, 2013. An amended version of this speech is online available under http://www.cm.sarawak.gov.my/en/media-centre/speeches/view/dun-sitting-may-2013-yab-pehin-sri-haji-abdul-taib-mahmud-winding-up-speech.

11 For information on the UN Declaration on the Rights of Indigenous Peoples see http://

undesadspd.org/
IndigenousPeoples/
Declarationon-
theRightsofIndi-
genousPeoples.aspx.

12 The NGO Malaysian
Election Observers
Network estimated that
in 2010, approximately
480,000 electors in
Sarawak were not
registered, 80% of
those in the rural
districts. Email
message from BK Ong,
Malaysian Election
Observers Network, to
the author, dated
March 11, 2011.

13 See also http://
treaties.un.org/Pages/
ViewDetails.
aspx?s-
rc=TREATY&mtdsg_
no=IV-4&chap-
ter=4&lang=en.

14 Monika Roth,
"Aufsichtsrechtliche
Vorgaben für Banken
und Art. 102 StGB: Ein
Diskussionsbeitrag zu
den beiden Alstom-
Entscheiden der
Bundesanwaltschaft,"
Jusletter (June 18,
2012); and Monika
Roth, "Compliance
darf weder Papiertiger
noch lahme Ente sein:
Zwei Alstom-
Entscheide in der
Schweiz," *ZRFC,* no. 4
(2012), 174 ff.

15 Schweizer Radio DRS,
"Bundesanwaltschaft

ermittelt gegen die
UBS," August 31,
2012, http://drs.srf.ch/
www/de/drs/
nachrichten/wirtschaft/
359851.bundesan-
waltschaft-ermittelt-
gegen-die-ubs.html.

16 Preamble to the United
Nations Convention
against Corruption,
Classified Compilation
of Swiss Federal Law,
SR 0.311.56.

17 Articles 17, 18, 19, and
20 of the United
Nations Convention
against Corruption,
Classified Compilation
of Swiss Federal Law,
SR 0.311.56.

18 Website of the FBI
Field Office in Seattle,
accessed June 10, 2013,
http://www.fbi.gov/
seattle/about-us/
what-we-investigate/
priorities.

19 Clare Rewcastle,
"Where is Farok
Majeed and How
Wealthy is Onn
Mahmud?," *Sarawak
Report*, March 18,
2011, http://www.
sarawakreport.
org/2011/03/where-
is-farok-majeed-and-
how-wealthy-is-onn-
mahmud.

20 Mark Baker "Tycoon
dodges millions in land
tax," *Age*, April 28,
2013, http://www.
theage.com.au/

national/tycoon-dodg-
es-millions-in-land-tax-
20130427-2ilmn.html.

21 Dev Kar and Sarah
Freitas, *Illicit
Financial Flows from
Developing Countries
2001–2010*, Global
Financial Integrity
(December 2012),
http://www.gfintegrity.
org/wp-content/
uploads/2014/05/
Illicit_Financial_
Flows_from_Develop-
ing_Countries_2001-
2010-HighRes.pdf.

22 IUCN Red List,
Presbytis chrysomelas,
accessed June 10, 2013,
http://www.iucnred-
list.org/de-
tails/39803/0.

23 SarVision, *Impact of oil
palm plantations on
peatland conversion in
Sarawak 2005–2010*,
Summary report
(January 25, 2011), 11
ff.; see also Global
Witness, *Sarawak's
Forests: Myths &
Reality* (May 2013).

24 See also http://wwf.
panda.org/what_we_
do/where_we_work/
borneo_forests/
borneo_rainforest_
conservation/
declaration.cfm.

25 Nigel Aw, "MACC
plays down Taib's
'naughty, dishonest'
remark," *Malaysiakini*,
June 29, 2013.

26 Letter from the Canadian Minister of Finance, James Flaherty, to the Bruno Manser Fund, dated September 14, 2011.

27 Letter from the Royal Canadian Mounted Police to the Bruno Manser Fund, dated July 26, 2011.

28 Letters from the Australian Government, Department of Foreign Affairs and Trade, to the Bruno Manser Fund, dated September 28, 2011, and from the Australian Federal Police to the Bruno Manser Fund, dated October 28, 2011.

29 Letter from the Foreign and Commonwealth Office to the Bruno Manser Fund, dated November 8, 2011.

30 Letter from John Harris, Director General, Jersey Financial Services Commission, to the Bruno Manser Fund, dated October 6, 2011.

31 Email from Sarah Merzbach, German Federal Ministry of Finance, Directorate VII A 3 (Payment systems, prevention of money laundering) Division VII, to the Bruno Manser Fund, dated September 5, 2011.

32 Letter by Hartmut Koschyk, parliamentary state secretary at the German Federal Ministry of Finance, to Dr. Thomas Gambke, member of the German *Bundestag*, dated March 8, 2012.

33 See also Interpol website on environmental crimes, http://www.interpol.int/Crime-areas/Environmental-crime/Environmental-Compliance-and-Enforcement-Committee/Pollution-Crime-Working-Group.

34 Letter from the Interpol General Secretariat, Office of Legal Affairs, to the Bruno Manser Fund, dated December 20, 2011.

35 Letter from the Bruno Manser Fund to the MACC (Malaysian Anti-Corruption Commission), the Attorney General of Malaysia, and the Inspector General of Police, December 13, 2011, http://www.stop-timber-corruption.org/resources/Taib_Arrest_Letter_1.pdf.

36 Goncalves Pereira et al., *Justice for Forests. Improving Criminal Justice Efforts to Combat Illegal Logging*, World Bank Study (Washington DC, 2012), vii.

37 Statement by David Higgins, Head of the Interpol Environmental Crime Programme, cited from Environment News Service, "Interpol Arrests 194 in Illegal Logging Sting," February 20, 2013.

38 United Nations Office on Drugs and Crime *Transnational Organized Crime in East Asia and the Pacific*, A Threat Assessment, April 2013.

39 Interpol, "INTERPOL launches Project LEAF to combat illegal logging worldwide," media release, June 5, 2012, http://www.interpol.int/News-and-media/News-media-releases/2012/N20120605Bis.

40 United Nations Office on Drugs and Crime, *Transnational Organized Crime*, April 2013, 90.

41 Ibid. 95.

42 FERN, *Forest Watch Special—VPA Update* (November 2012), 4.

43 For information on the EU's FLEGT process, the reader is referred to the excellent thematic

website of the British Royal Institute of International Affairs (Chatham House), www.illegal-logging. info.

44 See also ec.europa.eu/ environment/eutr2013/ index_de.htm.

45 Parliament of Australia, "Illegal Logging Prohibition Bill 2012," http://www.aph.gov. au/%20Parliamentary_ Business/%20 Bills_Legislation/%20 Bills_Search_Re- sults/%20Re- sult?bId=r4740.

46 Philip Shearman, Jane Bryan, and William F. Laurance, "Are we approaching 'peak timber' in the tropics?," *Biological Conservation* 151, no. 1 (2012), 17 ff.

47 Duncan Poore, and Thang Hooi Chiew, *Re- view of Progress towards the Year 2000 Objective*, International Tropical Timber Coun- cil (November 2000).

48 Jürgen Blaser et al., "Status of Tropical For- est Management 2011." ITTO Technical Series, no. 38, International Tropical Timber Organization (Yokohama, 2011); see also Swiss State Secretariat for Economic Affairs (SECO), "Das globale Engagement zum Schutz des Tropenwal- des muss gesteigert werden," press release, June 7, 2011, http:// www.seco.admin.ch/ aktuell/ 00277/01164/01980/ index.html?lang= de&msg-id=39477.

49 Minutes of the first ordinary general meeting of the Bruno Manser Fund (BMF) held on December 7, 1991, in Les Pommer- ats. Bruno Manser Fund archive, Basel.

50 "6 mln ha of planted forest by 2020," *Borneo Post*, March 26, 2013.

SARAWAK CHRONOLOGY

c. 40000 BCE
Human occupation of the Niah caves in Sarawak.

14th century
Earliest documentary reference to "Serawak".

1841
James Brooke granted governmental powers as "Rajah" of Sarawak by the Sultan of Brunei.

1855
The naturalist and explorer Alfred Russel Wallace formulated the "Sarawak law" of evolution.

1868
Death of the First White Rajah, James Brooke; succeeded by his nephew, Charles Brooke.

1894
Colonial officer Charles Hose wrote the first-ever description of the rainforest people of the "Punan" (Penan) of Sarawak.

1917
Charles Vyner Brooke became the Third Rajah of Sarawak.

1936
Abdul Taib Mahmud was born as the oldest child of a carpenter living near Miri.

1941
Japanese troops occupied Sarawak.

1946
One year after the end of the Second World War, Sarawak became a British Crown colony.

1951
Rodney Needham, who was later to become an Oxford professor, started work on his dissertation about the Penan.

1956
Taib studied law at the University of Adelaide.

1957
The Federation of Malaya (today's "Peninsular Malaysia") was granted independence from British colonial rule.

1962
Taib returned to Sarawak with his wife, Laila, and daughter, Jamilah.

1963
Sarawak became part of the state of Malaysia; Taib became a minister in Sarawak.

1966
Stephen Kalong Ningkan, the first chief minister of Sarawak, was removed from office.

1966
Peace treaty between Malaysia and Indonesia, end of three years of confrontation between the two countries.

1968
Taib began his ministerial career in the Malaysian federal government.

1969
Racial unrest in Peninsular Malaysia; Sarawak's chief minister, Tawi Sli, was ousted from power.

1970
Rahman Ya'kub, Taib's uncle, became chief minister of Sarawak.

1980
First organised protest by indigenous people against deforestation in Sarawak.

1981
Taib became chief minister of Sarawak, while his uncle assumed the office of governor.

1983
Incorporation of the Sakto real estate company in Canada by Onn Mahmud on behalf of his brother Taib.

1984
Bruno Manser arrived in Sarawak with the intention of living amongst the nomadic Penan in the virgin forest.

1985
Taib abolished the forestry ministry and assumed control of all logging concessions and plantation licences.

1986
The German magazine Geo published a report on Bruno Manser and the Penan resistance to deforestation.

1987
Incorporation of Taib's Sakti real estate company in California.

1987
Protests in the form of roadblocks against the logging industry by thousands of indigenous inhabitants.

1987
Power struggle between Taib and his uncle ("Ming Court affair").

1987
Arrest of more than a hundred members of the opposition as part of Operation Lalang ("Weeding-out Operation").

1989
More than 4,000 indigenous inhabitants joined in new blockades against deforestation.

1990
Bruno Manser's return to Switzerland.

1992
Indigenous lawyer Baru Bian set up his legal practice of Messrs Baru Bian in Kuching.

1993
The Taib family embarked upon its "reverse takeover" of the state construction company Cahya Mata Sarawak (CMS).

1993
300 police broke up the Penan blockade near Long Sebatu using tear gas; one child killed.

1994
Ross Boyert became director of Taib's Sakti real estate company in the USA.

1996
Incorporation of the Taibs' Ridgeford Properties company in London.

1997
Taib caused the failure of a plan supported by the World Bank to create a biosphere reserve for the Penan.

1998
The FBI became tenants in the Abraham Lincoln Building owned by Taib in Seattle.

2000
Bruno Manser disappeared without a trace in the Sarawak rainforest.

2001
The Taib family purchased Malaysia's fourth largest bank, RHB, for 1.8 billion ringgits (US$ 500 million).

2001
For the first time ever, the High Court of Sarawak and Sabah granted land rights over intact virgin forest to an indigenous community.

2004
UBS floated a government loan worth US$ 350 million on behalf of the Taib regime.

2005
Deutsche Bank floated a government loan worth US$ 600 million on behalf of the Taib regime.

2005
The civil court in the Swiss canton of Basel-Stadt declared Bruno Manser to be missing, presumed dead.

2006

Ross Boyert was dismissed by Sakti and replaced by Taib's son-in-law, Sean Murray.

2007

Credit Suisse underwrote the launch of the Malaysian Samling timber company on the Hong Kong stock exchange.

2008

The University of Adelaide named a plaza on its campus after Taib.

2010

Taib announced at an electoral event that "I have more money than I could ever spend in my life".

2010

Commissioning of Bakun Dam—described by Transparency International as a "monument of corruption".

2011

The Taib family had a financial interest in over 400 companies in 25 countries and offshore finance centres.

2011

Opening of an inquiry into Taib by the Malaysian anti-corruption commission (MACC).

2011

Goldman Sachs floated an initial financial-market loan worth US$ 800 million on behalf of the Taib regime.

2012

Swiss criminal proceedings against UBS on suspicion of laundering the proceeds of corruption from Malaysia.

2012

The Bruno Manser Fund estimated the Taib family's assets at US$ 20 billion.

2013

Protest by hundreds of indigenous people in Kuching against the Taib regime's plans for hydropower.

2013

Radio Free Sarawak received the Free Media Pioneer Award from the International Press Institute in Vienna.

2014

Taib resigned as chief minister after 33 years in office and was appointed governor of Sarawak.

ACKNOWLEDGEMENTS

Many people have helped to make this book possible—some of whom wish to remain anonymous. The author would like to thank in particular the Bruno Manser Fund, Salis Verlag, Bergli Books/Schwabe AG, and the following persons:

Annina Aeberli, Bruce Bailey, Baru Bian, Emeric Billard, Jürgen Blaser, Dominik Bucheli, Julien Coquentin, Wade Davis, Amy Dodds, Thomas Gierl, Marion Graber, André Gstettenhofer, Richard Harvell, Sally Holloway, Peter Kallang, Vernon Kedit, Christoph Lanz, Dorothee Lanz, Tracey Lauriault, Michael Leuenberger, Simon Kaelin, Ian Mackenzie, Joe Jengau Mela, Johanna Michel, Robert Middleton, Antoinette and Kaspar Müller, Tristan Needham, Harrison Ngau, John Palmer, Eseng Pege, Heini Pestalozzi, Peg Putt, Clare Rewcastle, John Rewcastle, Asti Roesle, Monika Roth, Eva Ruch, Datuk Salleh Jaffaruddin, Patrick Schär, Rolf Schenk, See Chee How, Eva Spehn, Lynette Tan, Felix Thomann, Daniela Trunk, Mutang Urud, Valentine Vogel, Eric Wakker, Jenny Weber, Rainer Weisshaidinger, Daniel Wildmann, Wee Aik Pang, Wong Meng Chuo, and Erwin Zbinden.

PHOTO CREDITS

INDEX